中等职业教育中餐烹饪专业系列教材

烹饪英语

第2版

主　编　杜　纲
副主编　杜茹薇

重庆大学出版社

内容提要

本书以烹饪常用词汇为主线，涉及10个方面的内容，包括蔬菜、水果、小吃、肉类、配料、烹饪用具、烹饪手法、干果和蜜饯、海鲜、饮料等，并配上相应的图片。全书共分10个单元，每个单元以上面其中一个方面的单词为主，配以日常用语及对话、重点单词用法、英文菜谱。每个单元后都有自我评价、生词表，并附有中文菜品名译成英文菜品名的常用方法、酒店各部门的英文名称及最常用礼貌用语，以方便学生运用。每个单元配有大量的练习以方便学生记忆和运用。针对学生不同的学习进度及学生将来工作的需求，在一些菜谱后增加了图片以便学生复述。修订后的本书，增加了二维码辅助教学，为学生提供了新的学习形式。

本书可作为中等职业教育中餐烹饪专业教材，也可作为烹饪和英语爱好者的参考书。

图书在版编目（CIP）数据

烹饪英语 / 杜纲主编. --2 版 . -- 重庆：重庆大学出版社，2022.6

中等职业教育中餐烹饪专业系列教材

ISBN 978-7-5624-6974-2

Ⅰ.①烹… Ⅱ.①杜… Ⅲ.①烹饪—英语—中等专业学校—教材 Ⅳ.①TS972.1

中国版本图书馆 CIP 数据核字（2021）第118134号

中等职业教育中餐烹饪专业系列教材

烹饪英语（第 2 版）

主 编 杜 纲

副主编 杜茹薇

责任编辑：沈 静 版式设计：沈 静

责任校对：刘志刚 责任印制：张 策

*

重庆大学出版社出版发行

出版人：饶帮华

社址：重庆市沙坪坝区大学城西路 21 号

邮编：401331

电话：（023）88617190 88617185（中小学）

传真：（023）88617186 88617166

网址：http://www.cqup.com.cn

邮箱：fxk@cqup.com.cn（营销中心）

全国新华书店经销

重庆升光电力印务有限公司印刷

*

开本：787mm × 1092mm 1/16 印张：6.75 字数：201 千

2013 年 1 月第 1 版 2022 年 6 月第 2 版 2022 年 6 月第 6 次印刷

印数：13 001—15 000

ISBN 978-7-5624-6974-2 定价：35.00 元

第2版前言

作为中等职业教育中餐烹饪专业系列教材之一的《烹饪英语》在各方帮助和支持下终于完稿了。本书主要适用于中职学生，同时也适合从事烹饪专业的人士学习。

本书本着“实用为主，够用为度”的原则，以中澳项目的科研成果为理念，着重强调将实际工作中常用的英语交流用语、专业术语、专业词汇提供给学生训练与记忆。同时让学生学会阅读英语菜谱及有关食品使用说明，以便在实际工作中加以运用。

本书最大的特点在于：将烹饪英语常用词按不同属性分为 10 个部分并配以大量的图片，同时以单词为核心组成 10 个单元，每个单元的内容都体现了本单元单词的引领作用，非常清晰，便于学生记忆。

第 2 版作了以下改动：

①增加了二维码辅助教学，为学生提供新的学习形式，课后学生可以反复学习。

②强化了字母及字母组合在单词中的主要读音。

③对各单元词汇加注了音标。

④替换了大量图片，使其更加清晰、易认。

⑤制作了全书的课件并给出了练习答案以供参考。

⑥对主要单词、对话和常用术语配备了英文录音，以便学生学习。

本书由重庆市旅游学校杜纲担任主编，重庆市旅游学校杜茹薇担任副主编。杜纲负责所有章节的修订。杜茹薇对所有图片进行了调整并制作了课件。

本书在编写过程中得到了有关专家的大力支持。同时得到了澳大利亚专家 Jeff 的指导。本书还参考了一些图片和资料，在此一并表示感谢！

编　者

2022 年 1 月

第1版前言

作为中等职业教育中餐烹饪专业系列教材之一的《烹饪英语》在各方帮助和支持下终于完稿了。本书主要适用于中职学生，同时也适合从事烹饪专业的人士学习。

本书本着“实用为主，够用为度”的原则，以中澳项目的科研成果为理念，着重强调将实际工作中常用的英语交流用语、专业术语、专业词汇提供给学生训练与记忆。同时，让学生学会阅读英语菜谱及有关食品使用说明，以便在实际工作中加以运用。

本书最大的特点在于：将烹饪英语常用词按不同属性分为10个部分并配以大量的图片，同时以单词为核心组成10个单元，每个单元的内容都体现了本单元单词的引领作用，非常清晰，便于学生记忆。

本书由杜纲担任主编，负责书稿的编写。李雪雯负责修改和完善Unit 5，Unit 6，Unit 7，蔡琳琳负责修改和完善Unit 10，张华负责修改和完善Unit 2，Unit 3，Unit 5，区畅负责修改和完善Unit 9。

本书在编写过程中得到了有关专家的大力支持，同时也得到了澳大利亚专家Jeff的指导。本书还参考了大量图片和资料，在此一并表示感谢！

编　者

2012年10月

Contents

UNIT ONE Vegetables

Look and say 1
Speak out 2
1. Learn the alphabet 2
2. Learn the ABC song 2
3. Learn the sounds of the letters in word 3
4. Everyday English 4
Dialogues 4
Greetings 4
Dialogue I 4
Dialogue II 5
Useful sentences with *cut* 5
More words about vegetables 6
Recipe 8
Green Beans with Chili Garlic Sauce 8
Assignment 9
Self-evaluation 9
Word list for Unit One 10

UNIT TWO Snacks

Look and say 12
Speak out 13
1. Learn the vowel 13
2. Learn the sounds of the letters in word 13
3. Everyday English 13
Dialogue 14
Where are you from? 14
Useful sentences with *remove* 15
More words about snacks 15

Recipe 17
Lobster Spring Roll 17
Assignment 18
Self-evaluation 19
Word list for Unit Two 19

UNIT THREE Meat

Look and say 21
Speak out 22
Everyday English 22
Dialogues 22
Receptions 22
Dialogue I 22
Dialogue II 23
Useful sentences with *steam* 24
More words about meat 24
Recipe 25
Beef Stew 25
Assignment 27
Self-evaluation 27
Word list for Unit Three 27

UNIT FOUR Fruits

Look and say 29
Speak out 30
Everyday English 30
Dialogue 30
Serving guests 30
Useful sentences with *pour* 31
More words about fruit 32
Recipe 33
Orange and Strawberry Salad 33
Assignment 34
Self-evaluation 35
Word list for Unit Four 35

UNIT FIVE Ingredients

Look and say 37
Speak out 38
1. Learn the numbers 38
2. Tell the following telephone numbers and room numbers 38
3. Everyday English 39
Dialogues 39
Paying the bill 39
Dialogue I 39
Dialogue II 40
Useful sentences with *carve* 41
More words about ingredients 41
Recipe 43
Winter Stew 43
Assignment 45
Self-evaluation 45
Word list for Unit Five 46

UNIT SIX Nuts and Sweetmeats

Look and say 48
Speak out 49
1. Learn the expressions of time 49
2. Tell and write the following time in English 49
3. Everyday English 50
Dialogues 50
Dialogue I 50
Dialogue II 51
Useful sentences with *keep* 52
More words about nuts and sweetmeats 52
Recipe 53
Diced Chicken with Chili Pepper 53
Assignment 55
Self-evaluation 55
Word list for Unit Six 56

UNIT SEVEN Drinks

Look and say 57
Speak out 58
1. Learn the words about colours 58
2. Make a short dialogue according to the example 58
3. Everyday English 59
Dialogues 60
What do you have for breakfast? 60
Dialogue I 60
Dialogue II 60
Useful sentences with *sprinkle* 61
More words about drinks 62
Recipe 64
Champagne Punch 64
Assignment 65
Self-evaluation 65
Word list for Unit Seven 65

UNIT EIGHT Seafood

Look and say 67
Dialogue 68
Thai Prawn Soup and Chrysanthemum Fish 68
Useful sentences 69
More words about seafood 69
Recipe 71
Marinated Grilled Shrimp 71
Assignment 73
Self-evaluation 73
Word list for Unit Eight 73

UNIT NINE Tableware and Cooking Tools

Look and say 75
Speak out 76
Everyday English 76
Dialogue 76
Recommending dishes 76

Useful sentences with *garnish* 77
More word about tableware and cooking tools 77
Instruction 79
Delicious Biscuit 79
Assignment 80
Self-evaluation 80
Word list for Unit Nine 81

UNIT TEN Cooking Terms

Look and say 83
Dialogue 84
Sweet and sour dishes 84
Useful sentences with *split* 84
More word about cooking terms 85
Instruction 88
Nice Soy Milk 88
Assignment 89
Self-evaluation 90
Word list for Unit Ten 90

Supplementary Materials

A. Methods to Express the Names of Dishes 92
B. Courtesy English 93
C. English Names for Departments in Hotel 93

参考文献 95

UNIT ONE

1 Vegetables

Look and say

1.________________

2.________________

3.________________

4.________________

5.________________

6.________________

7.________________

8.________________

9.________________

10.________________

11.________________

12.________________

Task I Find out the suitable pictures for the given words

carrot	turnip	leek	lettuce
cauliflower	onion	cucumber	eggplant
celery	potato	tomato	bean sprout

Speak out

1. Learn the alphabet

A B C D E F G H I J K L M N O P Q R S T U V W X Y Z

a b c d e f g h i j k l m n o p q r s t u v w x y z

2. Learn the ABC song

I CAN SAY MY ABC

Task II Recite the alphabet and sing the song

3. Learn the sounds of the letters in word

b [b] book
d [d] desk
j [dʒ] jeep
k [k] key
m [m] mop
n [n] name
p [p] pig
s [s] sit
t [t] tea
v [v] five
x [ks] six
z [z] zoo

c [s] lettuce
[k] cook
g [dʒ] porrige
[g] goat

h [h] hello
r [r] rice
w [w] we

y [j] yes
[ai] fly
[i] study

ch [tʃ] chest
sh [ʃ] fish
ph [f] phone
tch [tʃ] catch
tr [tr] try
dr [dr] dry
ts [ts] pots
ds [dz] hands
qu [kw] queen
ng [ŋ] thing

th [θ] thin
[ð] mother
wh[w] what
[h] who

Task III Pronounce the consonant letters

carrot	turnip	leek
lettuce	cauliflower	onion
eggplant	celery	potato
tomato	cucumber	mushroom
radish	asparagus	bamboo shoot
spinach	broccoli	white gourd
spring onion	basil	chili
tomato	pea	string bean
bean sprout	lotus root	pumpkin

4. Everyday English

- Hello.
- Good morning.
- Good afternoon.
- Good evening.
- What's your name, please?
- My name is Peter.
- Can I have your name please?
- I am Peter.
- Could you tell me your name please?
- Nice to meet you.
- How are you?
- I am fine, thanks.

Dialogues

Greetings

***Dialogue* I**

—Good morning, sir.

—Good morning. What's your name, please?

—My name is John.

—My name is Smith. Nice to meet you.

—Nice to meet you, too.

—How do you do?

—How do you do?

Task IV　Work in pairs

Fill in the blanks with correct words.

—Good afternoon, Sir.

—Good _______. What's your name, please?

—My _______ is Jim.

—My name is Karl. Nice to meet you.

—Nice to _______ you, too.

—How do you do?

—_______ do you do?

***Dialogue* II**

—Hello. How are you today?

—I am fine. And how are you?

—I am fine too. Thank you.

Task V　Work in pairs

Fill in the blanks with sentences.

—Hi. How are you ?

—______________________________

—Can I have your name, please?

—______________________________

—Nice to meet you.

—______________________________

Task VI　Role-play

You come to a new restaurant for work. The boss meets you for the first time.

Useful sentences with *cut*

Cut the pumpkin into 5 blocks.

Cut the leek into small sections.

Cut the potato into 5 pieces.

Cut the mushrooms into small sections and stir-fry until done.

Cut down the tomato to the skin but leave the skin intact.

Cut the scallion into shreds.

Cut the spring onion into cubes.

Cut the ginger into 6 slices.

Task VII Discussion

Discuss how to say the following sentences in English.

把莲藕切成两块。

把这个冬瓜切到皮，不要切断。

切开南瓜，去籽。

把大蒜切片。

More words about vegetables

mushroom

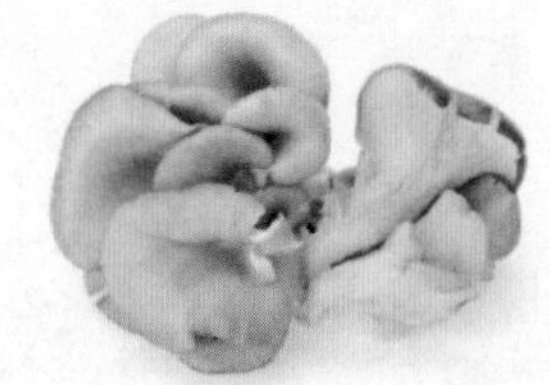
oyster mushroom

shitake mushroom

enoki mushroom

radish

asparagus

bamboo shoot

spinach

broccoli

spring onion
basil
chili
taro
capsicum
French bean
pea
green bean
lotus root
pumpkin
white gourd
bitter melon
corn
bean curd
yam

cabbage

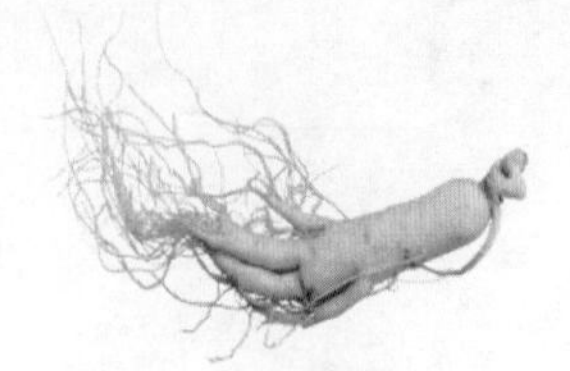
ginseng

Chinese cabbage

Task VIII Fill in the blanks with correct letters

mu____room	oys____ mushroom	shit____ke mushroom
____noki mushroom	r____dish	asp____rag____s
bamb____shoot	l____k	spina____
br____ccol____	spring ____nion	ba____il
t____ro	____ili	c____psicum
sc____llion	tom____to	French b____n
p____	s____ing bean	bean spr____t
lot____s r____t	c____c____mber	p____mpk____n
white g____d	bi____er melon	c____n
bean c____d	g____nseng	on____n
c____rrot	t____nip	lett____ce
c____liflower	egg pl____nt	c____lery
pot____to		

Recipe

Green Beans with Chili Garlic Sauce

- 3/4 pound green beans, trimmed
- 2 teaspoons olive oil
- 1/4 small red onion, thinly sliced (about 1/2 cup)
- 1/2 medium red bell pepper, thinly sliced (about 1/2 cup)

- 1/4 pound shitake mushrooms, sliced
- 1 clove garlic, minced
- 1/2 teaspoon red chili garlic sauce

Bring a large pot of water to a boil over high heat.

Add the green beans to the water. Cook the beans for about 4 minutes, or until they are bright green and still slightly crisp. Meanwhile, prepare a large bowl of ice water.

Drain the green beans in colander and quickly place them in the ice water to stop the cooking process. Drain the beans again in a colander before proceeding to the next step.

Heat the olive oil in a large skillet over high heat. Add the onion, pepper, and mushrooms, and cook for about 3 minutes, stirring constantly, or until the vegetables begin to brown slightly.

Add the green beans, garlic, salt, and pepper to the skillet. Cook for 30 seconds, or until the garlic is fragrant.

Add the red chili garlic sauce to the skillet and stir to coat the vegetables well. Remove from heat and serve.

Task IX　True or false

1. There are six vegatables in the dish. (　　)
2. This is a Chinese-style dish. (　　)
3. Deep-fly the beans for two minutes. (　　)
4. Heat the olive oil in a big skillet over high heat. (　　)

Assignment

1. Recite the alphabet and the rules of sounds.
2. Make sentences with the word *cut*.
3. Practise the dialogue with your partner.
4. Recite the words on vegetables.
5. Translate the passage into Chinese.

Self-evaluation

1. I can sing the alphabet song and remember the letters. (　　)
2. I know how to pronounce the words. (　　)
3. I can recite the dialogues. (　　)
4. I can use the word *cut* correctly. (　　)
5. I can understand the passage. (　　)

6. I can pronounce the words and remember them. (　　)
7. My questions are ________________________________.

Word list for Unit One

carrot [ˈkærət] 胡萝卜
turnip [ˈtə:nip] 芜菁
leek [li:k] 韭菜
lettuce [ˈletis] 莴苣
cauliflower [ˈkɔliflauə] 花椰菜
onion [ˈʌnjən] 洋葱
eggplant [ˈegplænt] 茄子
celery [ˈseləri] 芹菜
potato [pəˈteitəu] 土豆
scallion [ˈskæljən] 葱
mushroom [ˈmʌʃru:m] 蘑菇
oyster mushroom [ˈɔistəˈmʌʃru:m] 平菇
shitake mushroom [ʃiˈtɑ:kiˈmʌʃru:m] 香菇
radish [ˈrædiʃ] 小萝卜
bamboo shoot [ˌbæmˈbu: ʃu:t] 竹笋
spinach [ˈspinitʃ] 菠菜
spring onion [ˌspriŋ ˈʌnjən] 小洋葱
taro [ˈtɑ:rəu] 芋头
bean curd [ˈbi:n kə:d] 豆腐
French bean [ˌfrentʃ ˈbi:n] 四季豆
string bean [ˌstriŋ ˈbi:n] 青豆
lotus root [ˈləutəs ru:t] 莲藕
pumpkin [ˈpʌmpkin] 南瓜
bitter melon [ˈbitə ˈmelən] 苦瓜
yam [jæm] 山药
pound [paund] 磅
olive oil [ˈɑ:liv ɔil] 橄榄油
teaspoon [ˈti:spu:n] 茶匙
clove [kləuv] 丁香
sauce [sɔ:s] 汁
cook [kuk] 烹调
slightly [ˈslaitli] 淡淡的
prepare [priˈpɛə] 准备
colander [ˈkʌləndə] 过滤器
proceeding [prəˈsi:diŋ] 进行过程中
stirring [ˈstə:riŋ] 搅拌
fragrant [ˈfreigrənt] 芳香的
enoki mushroom [iˈnəukiˈmʌʃru:m] 金针菇
asparagus [əˈspærəgəs] 芦笋
capsicum [ˈkæpsikəm] 辣椒、甜椒
broccoli [ˈbrɑ:kəli] 西兰花
basil ['beizl] 罗勒属植物
chili [ˈtʃili] 辣椒
tomato [təˈmeitəu] 番茄
pea [pi:] 豌豆
bean sprout [bi:n spraut] 豆芽
cucumber [ˈkju:kʌmbə] 黄瓜
white gourd [wait guəd] 冬瓜
corn [kɔ:n] 玉米
ginseng [ˈdʒinseŋ] 人参
trimmed [trimd] 修整过的，干净的
medium [ˈmi:diəm] 中等的
thinly [ˈθinli] 薄的
sliced [slaist] 切成片的
high heat [hai hi:t] 大火
meanwhile [ˈmi:nwail] 同时
crisp [krisp] 脆
drain [drein] 过滤
process [ˈprəuses] 过程
skillet [ˈskilit] 长柄锅
constantly [ˈkɔnstəntli] 不停地
block [blɔk] 块
section [ˈsekʃn] 段
stir-fry [ˈstə: frai] 炒

skin [skin] 皮
intact [inˈtækt] 完整的
shred [ʃred] 丝
slice [slais] 片
ginger [ˈdʒindʒə] 姜
cut [kʌt] 切

Chinese for the useful sentences.
将南瓜切成五大块。
将韭菜切成小节。
把土豆切成五片。
把蘑菇切成小节，炒熟。
将番茄切到皮但不切断。
把葱切成丝。
把小洋葱切成小方丁。
把姜切成六片。

Bird Singing

UNIT TWO

Snacks

Look and say

1. ________________

2. ________________

3. ________________

4. ________________

5. ________________

6. ________________

7. ________________

8. ________________

9. ________________

10. ________________

11. ________________

12. ________________

Task I Find out the suitable pictures for the given words

pizza hamburger sandwich hot pot
bread spring rolls egg salad
dessert dumpling noodle barbecued pork

Speak out

1. Learn the vowel

Aa Ee Ii Oo Uu

The open syllable:	The closed syllable:
A [ei] cake face plate name	[æ] bag map hat can
E [i:] me she we he	[e] ten hen egg let
I [ai] bike nice five like	[i] sit lit ship lip
O [əu] note hope hole bone	[ɔ] pot hot not shop
U [ju:] use cute refuse excuse	[ʌ] cut mug such cut

2. Learn the sounds of the letters in words

ar [ɑ:] far star	er [ə:] term teacher	ir [ə:] thirty	ur [ə:] turn
oa [əu] coat	oi [ɔi] oil	oo [u:] zoo too	ou [au] house
ow [au] cow	ee [i:] bee	eer [iə] beer	ue [u:] blue
ear [iə] ear	air [ɛə] air chair	ie [i:] believe	ay [ei] stay
ai [ei] train	aw [ɔ:] saw	ure [ə] picture	are [ɛə] prepare
or [ɔ:] port	or [ə] doctor	ea [e] bread	ea [i:] bean

Task II Try to pronounce the following words

cake face home go me she sheep student menu bike nice
mop bag ten pig cut boss map hen it hut
tail say sea meet soil boy tear beer shark per sir sort turn worker doctor
mouse how share draw book cool clue head curd pea

3. Everyday English

- How's everything with you?

- How are you doing?
- I am good.
- Can I help you?
- Where are you from?
- I'm from Chongqing.
- What do you like?
- How old are you?
- What's your job?
- I like cooking.
- I'm sixteen years old.
- I'm a cook.

Dialogue

Where are you from?

—Good morning, sir.
—Good morning. Can I help you?
—Yes, I'd like to work here.
—OK. Where are you from?
—I'm from Chongqing.
—How old are you?
—I am sixteen.
—What's your job?
—I am a cook.
—Well, please fill out the form.
—OK. Thank you.

Task III　Finish the dialogue with correct sentences

—Good morning, sir.
—Good morning. ______________________________
—Yes, I'd like to work here.
—OK. Where are you from?
—______________________________
—How old are you?
—______________________________
—What's your job?
—______________________________

—Well, please fill out the form.

—________________________________

Task IV Role-play

You want to find a job in a restaurant. The manager asks some questions and you answer him.

Useful sentences with *remove*

Remove the cream.
Remove the white from the egg.
Remove the mushrooms' stems.
Remove the froth from the stock.
Remove the apple pieces to an oiled plate.

Task V Discuss how to say the sentences in English

把西红柿的皮去掉。
去掉鱼尾。
去掉葱梗。
去掉水中的泡沫。

More words about snacks

starter

fried rice

pasta

soup

skewer

congee

crispy rice　prawn cracker　sticky rice

glutinous rice　buffet　à la carte

cereal　cornflakes　oatmeal

mashed potatoes　hand-made noodle　rice noodle

wonton　cake　pancake

Task VI　Fill in the blanks with correct letters

st____ter　　s____p　　past____

fr____d rice
crisp____ rice
sti____y rice
b____ffet
c____nflakes
pizz____
spring r____lls
s____lad
b____bec____ed pork
r____ce noodle

sk____er
pr____n
glutin____s rice
à la c____te
____t meal
sandwi____
br____d
sag____
dess____t
w____nton

cong____
table d'____te
hamb____ger
cer____l
ma____ed potatoes
hot p____t
____gg
sk____wer
hand-made n____dle
d____mpling

Recipe

Lobster Spring Roll

Ingredients:

- 2 (1/2-pound) lobsters
- 2 tablespoons sesame oil
- 2 teaspoons minced ginger
- 2 teaspoons minced garlic
- 2 small red Thai chili
- 1/2 pound cabbage, shredded
- 1 large carrot, peeled
- 1/2 cup sliced green onions
- 1 teaspoon salt
- 3/4 teaspoon ground white pepper
- 1 large egg
- 1 tablespoon water
- 16 spring roll wrappers
- peanut oil

Directions:

Fill a large stockpot with water and bring to a boil. Add lobsters and cook for 4 minutes. Remove the lobsters from the pot and plunge into an ice water bath, stirring until cool.

Remove the tail and claw meat of the lobsters' and dice into 1/2-inch pieces and set aside.

Heat the sesame oil in a wok over high heat just until smoking. Add the ginger, garlic, and chili and stir-fry for 1 minute, until fragrant. Add the diced lobster meat and cook for 1 minute. Turn the heat off and add the cabbage, carrots, and green onions.

Whisk the egg and water in a small bowl. Place 1/4 cup of the lobster mixture on each wrapper and fold in each side.

Fill a wok or deep pot halfway with peanut oil and heat. Drop the spring rolls, one by one, into the pot. Deep-fry until golden brown, about 2 minutes. Remove and drain on a paper towel-lined platter.

Task VII Answer the questions

1. How many ingredients are needed for Lobster Spring Roll?

2. How many vegetables are there in the snack?

3. Do we need peanut oil in this dish?

4. Can you make spring rolls?

Assignment

1. Recite the rules of sounds.
2. Make sentences with the word *remove*.
3. Practise the dialogue with your partner.

4. Recite the words on snacks.
5. Translate the passage into Chinese.

Self-evaluation

1. I can pronounce the words according to the rules. (   )
2. I can recite the dialogue. (   )
3. I can use the word *remove* correctly. (   )
4. I can understand the passage. (   )
5. I can remember all the words. (   )
6. My questions are ________________________________.

Word list for Unit Two

pizza [ˈpiːtsə] 比萨
sandwich [ˈsænwitʃ] 三明治
spring rolls [spriŋ rəulz] 春卷
egg [eg] 蛋
sago [ˈseigəu] 西米
starter [ˈstɑːtə] 餐前小吃
soup [suːp] 汤
fried rice [fraid rais] 炒饭
congee [kɔndʒiː] 粥
prawn cracker [ˌprɔːnˈkrækə] 虾片
sticky rice [ˈstiki rais] 糯米
buffet [ˈbufei] 自助餐
cereal [ˈsiəriəl] 谷类
oatmeal [ˈəutmiːl] 麦片
hand-made noodle [hænd meid ˈnuːdl] 手拉面
wonton [ˈwɑːntɑːn] 馄饨
lobster [ˈlɔbstə] 龙虾
stem [stem] 梗
teaspoon [ˈtiːspuːn] 茶匙
wrapper [ˈræpə] 面皮
minced [minst] 磨碎的
shredded [ˈʃredid] 切成丝的
sliced [slaist] 切成片的
peanut oil [ˈpiːnʌt ɔil] 花生油
hamburger [ˈhæmbəːgə] 汉堡
hot pot [hɔt pɔt] 火锅
bread [bred] 面包
salad [ˈsæləd] 沙拉
barbecued pork [ˈbɑːbikjuːd pɔːk] 叉烧
dessert [diˈzəːt] 甜品
pasta [ˈpɑːstə] 意大利面食
skewer [ˈskjuːə] 肉串
crispy rice [ˈkrispi rais] 锅巴
table d'hôte [ˈtɑːbəl dəut] 套饭
glutinous rice [ˈgluːtənəs rais] 糯米
à la carte [ˌæləˈkɑːt] 零点餐
cornflakes [ˈkɔːnfleiks] 玉米片
mashed potatoes [mæʃt pəˈteitəuz] 土豆泥
rice noodle [raisˈnuːdl] 米粉
dumpling [ˈdʌmpliŋ] 水饺
froth [frɔθ] 泡沫
tablespoon [ˈteiblspuːn] 大汤匙
pound [paund] 磅
sesame oil [ˈsesəmi ɔil] 芝麻油
Thai chili [tai ˈtʃili] 泰椒
peeled [piːld] 去皮的
ground [graund] 磨成粉的
stockpot [ˈstɔkpɔt] 汤料锅

plunge [plʌndʒ] 陷入
wok [wɔk] 炒菜锅
stir-fry [ˈstəː frai] 炒
turn off [ˈtəːn ɔf] 关掉
bowl [bəul] 碗
drop [drɔp] 放
drain [drein] 滤
cake [keik] 蛋糕
dish [diʃ] 一道菜
stirring [ˈstəːriŋ] 搅拌
high heat [hai hiːt] 大火
fragrant [ˈfreigrənt] 芬香的
whisk [wisk] 搅动
fold [fəuld] 折叠
deep-fry [ˌdiːp ˈfrai] 炸
towel [ˈtauəl] 纸巾
pancake [ˈpænkeik] 烙饼
remove [riˈmuːv] 去掉

Chinese for the useful sentences.
去掉奶油。
去掉蛋清。
去掉蘑菇梗。
去掉汤中的泡沫。
把苹果块放到油盘上。

Beef Noodles

UNIT THREE

3 Meat

Look and say

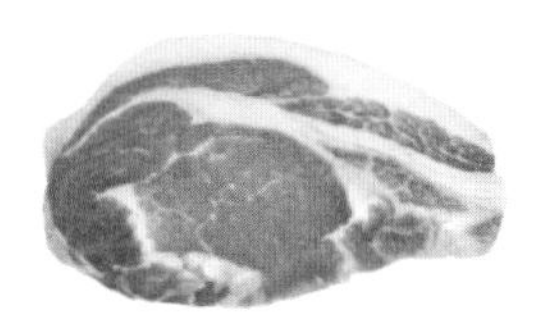

1.________________

2.________________

3.________________

4.________________

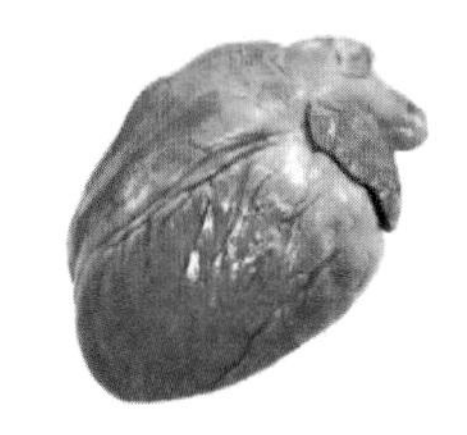

5.________________

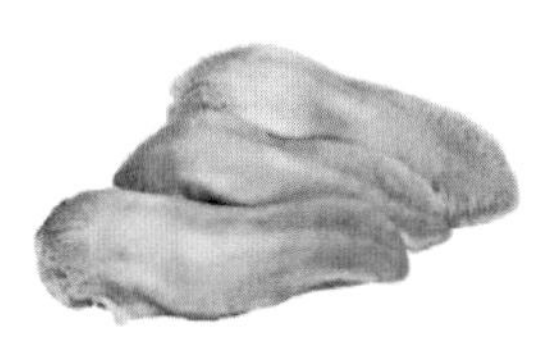

6.________________

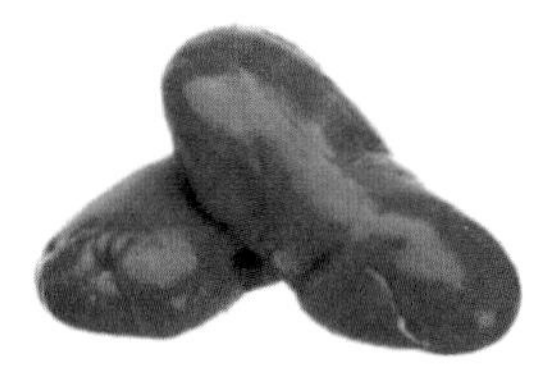

7.________________

8.________________

9.________________

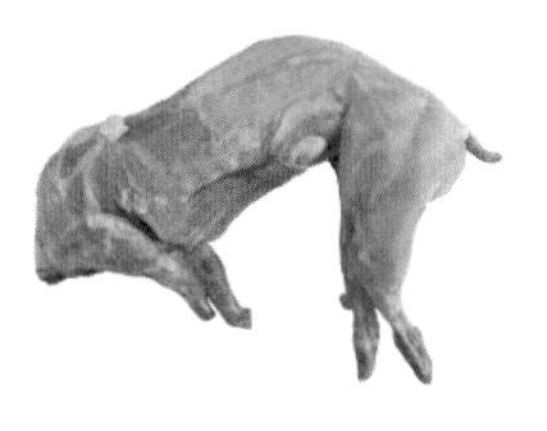

10.________________

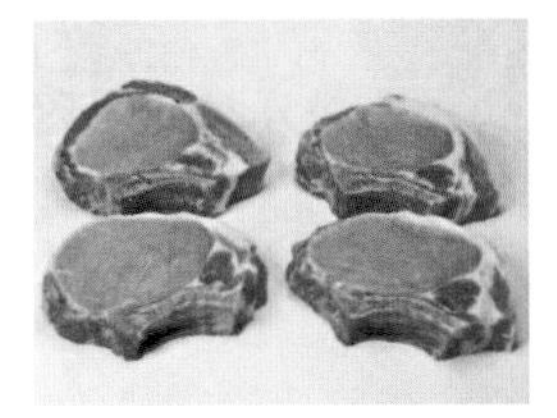

11.________________

12.________________

Task I Find out the suitable pictures for the given words

chicken	duck	pork chop	kidney
lamb	rabbit	steak	liver
tongue	heart	beef	pork

Speak out

Everyday English

- Excuse me, where's the kitchen please?
- It's over there.
- Welcome to our restaurant.
- It's on the second floor.
- Have you got a reservation?
- This way, please.
- Yes, of course.
- Sorry.
- Take the lift on the right.
- How can we get there?
- Could you please wait for a moment?

Task II Read the sentences fluently

Dialogues

Receptions

***Dialogue* I**

—Good evening, ladies and gentlemen. Have you got a reservation?

—Yes, of course.

—Can I have the name, please?

—Yes, Mr. Lee.

—Mr. Lee. The Rose Room on the second floor.

—How can we get there?

—Take the lift on the right, please.

—Thank you.

Task III Fill in the blanks with correct words

—Good evening, sir. Have you got _______?

—Yes, _______.

—Can I _______ the name, please?

—Yes, _______.

—Mr. Lee. The Rose Room _______ the first floor.

—How can we _______ there?

—_______ the lift on the left, please.

—Thank you.

***Dialogue* II**

—Good evening, madam. Can I help you?

—Yes, I'd like to have a table for four.

—Have you got a reservation?

—No, we've just come here.

—Sorry, our room is full now. Could you please wait for a moment?

—OK.

—Please take this seat.

—Thank you.

—It's my pleasure.

Task IV Finish the dialogue with correct sentences

—Good evening, sir. _______________________________

—Yes, I'd like to have a table for two.

—_______________________________

—No, we've just come here.

—_______________________________

—Could you please wait for a moment?

—OK.

—_______________________________

—Thank you.

—It's my pleasure.

Task V Role-play

A guest comes to your restaurant. But he has not reserved a table yet. You manage a table for him.

Useful sentences with *steam*

Steam the chicken for half an hour.
Steam the fish over high heat.
Steam the meat over low heat.
Steam it until well-done.

Task VI Discussion

Discuss how to say the sentences in English.
蒸半小时饭。
用大火蒸猪肉。
文火蒸蛋。
蒸得适中就行。

More words about meat

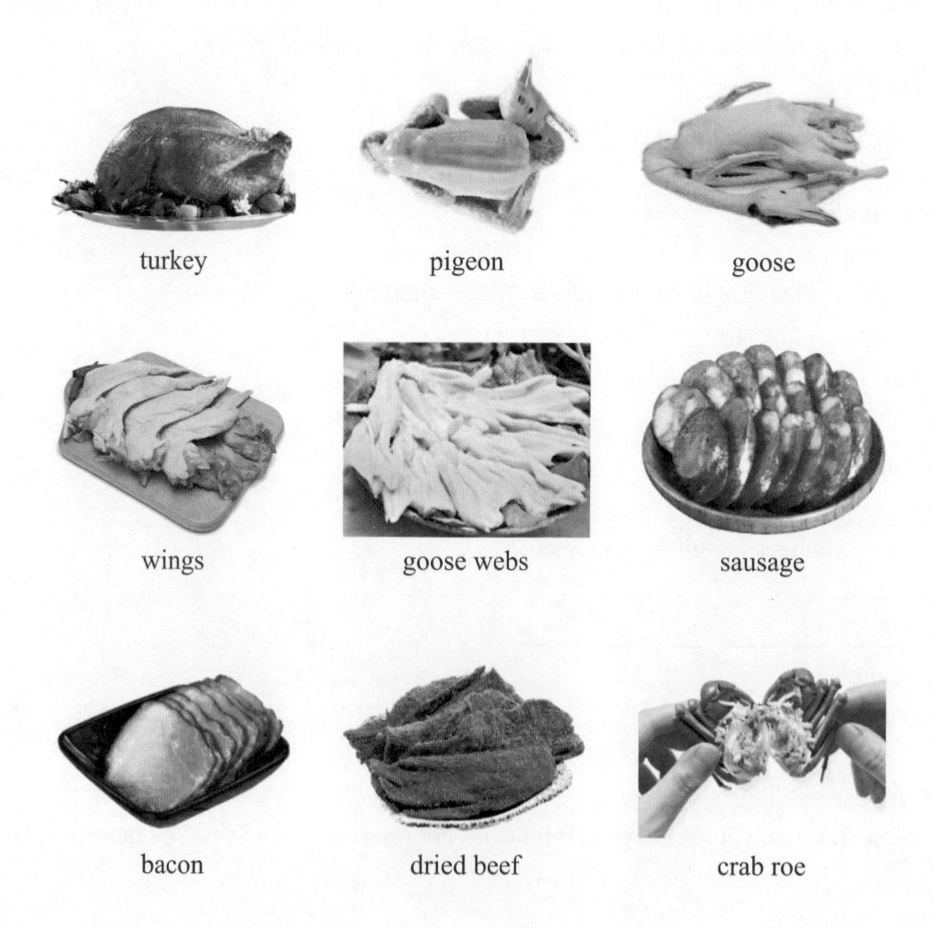

turkey pigeon goose
wings goose webs sausage
bacon dried beef crab roe

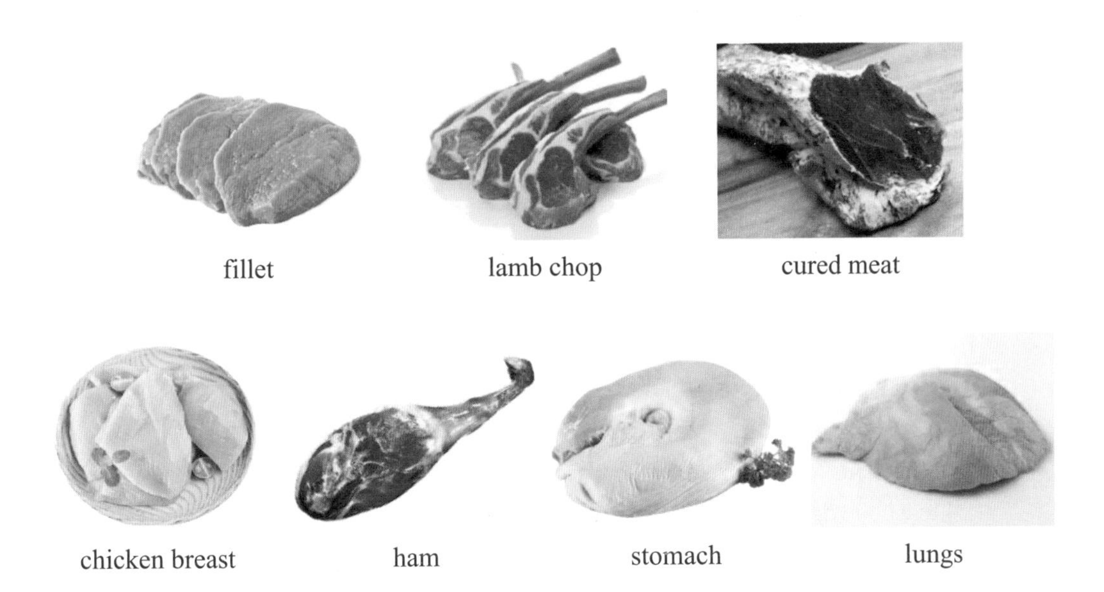

fillet lamb chop cured meat

chicken breast ham stomach lungs

Task VII Fill in the blanks with correct letters

chi____en	d____ck	p____k
kidn____	l____mb	r____bbit
l____ver	t____ngue	h____t
b____f	pig____n	st____k
t____key	g____se	goose w____bs
w____ngs	s____sage	b____con
____ied duck	crabr____	fill____t
pork ____op	cured m____t	bac____n
lamb ch____p	chicken br____st	h____m
st____mach	l____ngs	

Recipe

Beef Stew

Ingredients:

- 2 tablespoons olive oil
- 2 pounds beef stew meat, cut into 1-inch pieces
- 1 1/2 teaspoons salt
- 1 teaspoon essence, recipe follows
- 3/4 teaspoon cracked black pepper
- 2 tablespoons unsalted butter
- 1/2 pound button mushrooms, thinly sliced
- 3 tablespoons all-purpose flour
- 3 cups beef stock
- 2 tablespoons tomato paste
- 1/4 teaspoon dried thyme
- 1/4 teaspoon dried oregano
- 1/4 teaspoon dried basil
- 1/8 teaspoon ground allspice
- 1 pound small (golf ball size) new potatoes, quartered
- 1 cup diced carrots
- 1 cup frozen pearl onions, thawed
- 1/2 cup frozen green peas, thawed
- 1 tablespoon chopped fresh parsley leaves

Directions:

Season the beef with 1 teaspoon of the salt, 1 teaspoon essence and 1/2 teaspoon of the black pepper. Set a 12-inch saute pan over medium-high heat.

Add 1 tablespoon of the olive oil to the pan and sear the beef (in 2 batches) in the sauté pan for about 2 or 3 minutes per side.

Add the butter, mushrooms, flour, veal stock, tomato paste, herbs, spices, and browned meat to a slow cooker. Cover the slow cooker and set the temperature to high. Cook for 1 hour. Add the potatoes and carrots and continue to cook the stew for another 7 hours. During the last hour of cooking, add the pearl onions and replace the lid. Once the stew is cooked, stir in the peas and parsley and serve immediately.

Task VIII True or false

1. Season the beef with a tablespoon of salt. (　　)
2. Put some salad oil into the pan. (　　)
3. Add potatoes and carrots to cook for seven hours. (　　)
4. When the stew is cooked, add some peas and parsley. (　　)

Assignment

1. Recite the words.
2. Make sentences with the word *steam*.
3. Practise the dialogue with your partner.
4. Recite the words on vegetables.
5. Translate the passage into Chinese.

Self-evaluation

1. I can pronounce the w
2. I can speak out the w ing to the pictures. ( )
3. I can recite the dialogue. (
4. I can use the word *steam* correctly. ( )
5. I can understand the passage. ( )
6. My questions are ______________________________.

Word list for Unit Three

chicken [ˈtʃikin] 鸡肉
pork [pɔ:k] 猪肉
kidney [ˈkidni] 腰子
rabbit [ˈræbit] 兔肉
liver [ˈlivə] 肝
heart [hɑ:t] 心
turkey [ˈtə:ki] 火鸡
pigeon [ˈpidʒin] 鸽肉
goose webs [gu:s webz] 鹅掌
sausage [ˈsɔ:sidʒ] 香肠
dried beef [draid bi:f] 牛肉干
fillet [ˈfilit] 里脊；肉片
cured meat [kjuəd mi:t] 腌肉
chicken breast [ˈtʃikin brest] 鸡脯
stomach [ˈstʌmək] (烹饪原料)肚子
olive oil [ˈɑ:ɔliv ɔil] 橄榄油
cracked black [krækt blæk] 黑胡椒碎
duck [dʌk] 鸭肉
chop [tʃɔp] 排骨
lamb [læm] 羊肉
steak [steik] 牛排
tongue [tʌŋ] 舌
beef [bi:f] 牛肉
goose [gu:s] 鹅肉
lungs [lʌŋz] 肺
wings [wiŋz] 翅膀
bacon [ˈbeikən] 咸肉、熏肉
crab roe [kræb rəu] 蟹黄
pork chop [pɔ:k tʃɔp] 猪排
lamb chop [læm tʃɔp] 羊排
ham [hæm] 火腿
stew [stju:] 炖
essence [ˈesəns] 植物精华液
pepper [ˈpepə] 胡椒

butter [ˈbʌtə] 黄油
flour [ˈflauə] 面粉
thyme [taim] 百里香(草)
ground allspice [graund ˈɔːlspais] 多香果粉
thawed [θɔːd] 解冻
temperature [ˈtemprətʃə] 温度
medium-high heat [ˈmiːdiəm hai hiːt] 中大火
herbs [həːbz] 香草
replace [riˈpleis] 替换
immediately [iˈmiːdiətli] 马上
dried [draid] 干的
frozen [ˈfrəuzn] 冰冻的
stock [stɔk] 高汤
all-purpose [ˌɔːl ˈpəːpəs] 通用的
tomato paste [təˈmeiːtəu peist] 番茄酱
oregano [əˈregənou] 牛至
peel [piːl] 去皮
parsley leaves [ˈpɑːsli liːvz] 香菜叶
season [ˈsiːzn] 调味
sear [siə] 烧焦
spice [spais] 香料
lid [lid] 盖子
recipe [ˈresəpi] 食谱
unsalted [ʌnˈsɔːltid] 无盐的
chopped [tʃɔpt] 切碎的

Chinese for the useful sentences.
蒸半小时鸡。
用大火蒸鱼。
文火蒸肉。
蒸得适中就行。

Double Cooked Pork Slices

UNIT FOUR

4

Fruits

Look and say

1.________________

2.________________

3.________________

4.________________

5.________________

6.________________

7.________________

8.________________

9.________________

10.________________

11.________________

12.________________

Task I Find out the suitable pictures for the given words

orange	apple	grape	banana
pear	kiwi fruit	litchi	lemon
date	peach	watermelon	pomegranate

Speak out

Everyday English

- What would you like to have?
- Do you like anything to drink?
- What's the soup made of?
- It's made of eggs and tomatoes.
- Be seated, please.
- Anything else?

Dialogue

Serving guests

—Be seated, please.

—Thank you.

—Here's the menu.

—(a moment later)

—Are you ready to order now?

—Yes.

—What would you like to have?

—I'd like a table d'hôte. What is the soup made of?

—It's made of tomatoes and eggs. Do you like anything to drink?

—An orange juice.

—Anything else?

—No, thanks.

—A table d'hôte and an orange juice. Is that right?

—Yes, that's right.

—Just a moment, please. I'll bring them to you soon.

Task II Finish the dialogue with correct sentences

—______________________________

—Thank you.

—________________________________

—(a moment later)

—________________________________

—Yes.

—________________________________

—I'd like a table d'hôte. What is the soup made of?

—________________________________. Do you like anything to drink?

—A bottle of beer.

—________________________________

—No, thanks.

—A table d'hôte and a bottle of beer. ________________________________

—Yes, that's right.

—Just a moment, please. I'll bring them to you soon.

Task III Role-play

Two guests come to your restaurant. They want to have à la carte. You recommend some for them.

Useful sentences with *pour*

Pour in 100g peanut oil.

Pour in some grapes.

Pour in some strawberry.

Pour in some watermelon juice.

Pour in some water chestnuts.

Task IV Discussion

Discuss how to say the sentences in English.

放半斤樱桃。

放几滴橘子汁。

放些大枣。

放些香蕉。

放些葡萄汁。

More words about fruit

persimmon	water chestnut	loquat

water caltrop	muskmelon	fig

Task V Fill in the blanks with correct letters

or____nge	a____le	gr____pe
ban____na	pine____pple	p___r
str____berry	waterm____lon	kiwi fr____t
lich____	ol____ve	long____n
m____ngo	bl____berry	coc__nut
l____mon	pl____m	p_____ch
sug____cane	____erry	d____ria
tang____rine	kumq____t	mand____rin
apr____cot	cr____b apple	p____simmon
water ch____stnut	lo____at	d____te
water c____trop	pomeg____nate	f____g
m____sk melon		

Recipe

Orange and Strawberry Salad

Preparation time: 10 minutes, and refrigerate for 1 hour

(Serves 6)

8 oz (225 g) strawberries

6 oranges

rind of 2 oranges

freshly squeezed orange juice mint

Wash the strawberries. Drain and place them in a serving bowl. Grate the orange rind over the strawberries. Remove the pith from the oranges and then slice them. Arrange them on top of the strawberries. Add a little orange juice and decorate the salad with sprigs of mint. Cover and refrigerate for an hour before serving.

Task VI True or false

1. Orange and strawberry salad is a kind of Chinese salad. (　　)
2. There are five oranges and six strawberries in the salad. (　　)
3. We don't need any orange juice. (　　)
4. Before serving, add some orange juice and decorate the salad with sprigs of mint. (　　)

Task VII Talk about the above dish according to the pictures

Assignment

1. Copy the words and recite them.
2. Make sentences with the word *pour*.
3. Practise the dialogue with your partner.
4. Write a short dialogue.
5. Translate the passage into Chinese.

Self-evaluation

1. I can pronounce and remember the words. ( )
2. I can understand the passage. ( )
3. I can recite the dialogue. ( )
4. I can use the word *pour* correctly. ( )
5. My questions are ______________________.

Word list for Unit Four

orange [ˈɔrindʒ] 橙子
grape [greip] 葡萄
pineapple [ˈpainæpl] 菠萝
strawberry [ˈstrɔːberi] 草莓
kiwi fruit [kiːwiː fˈruːt] 猕猴桃
olive [ˈɑːliv] 橄榄
mango [ˈmæŋgəu] 杧果
coconut [ˈkəukənʌt] 椰子
plum [plʌm] 李子
durian [ˈduəriən] 榴莲
apricot [ˈæprikɔt] 杏
persimmon [pəˈsimən] 柿子
sugar cane [ˈʃugə kein] 甘蔗
date [deit] 枣
pomegranate [ˈpɔmigrænit] 石榴
muskmelon [ˈmʌskˌmelən] 香瓜
preparation [ˌprepəˈreiʃn] 准备
refrigerate [riˈfridʒəreit] 冷冻
rind [raind] 外皮
squeeze [skwiːz] 榨取, 挤出
grate [greit] 压碎
arrange [əˈreindʒ] 排列
sprig [sprig] 带叶小枝
apple [ˈæpl] 苹果
banana [bəˈnɑːnə] 香蕉
pear [pɛə] 梨
watermelon [ˈwɔːtəmelən] 西瓜
litchi [ˈlaitʃiː] 荔枝
longan [ˈlɔŋgən] 龙眼
blueberry [ˈbluːberi] 蓝莓
lemon [ˈlemən] 柠檬
peach [piːtʃ] 桃子
cherry [ˈtʃeri] 樱桃
tangerine [ˌtændʒəˈriːn] 橘子
crab apple [ˈkræb æpl] 山楂
water chestnut [ˈwɔːtə tʃesnʌt] 马蹄
loquat [ˈloukwɑːt] 枇杷
water caltrop [ˈwɔːtə ˈkætrəp] 菱角
fig [fig] 无花果
oz [ɔz] 盎司
freshly [ˈfreʃli] 新近
mint [mint] 薄荷
pith [piθ] 橙子等外皮之下的海绵层
decorate [ˈdekəreit] 装饰
pour [pɔː] 倒

Chinese for the useful sentences.
放100克花生油。
放些葡萄。
放点草莓。
放些西瓜汁。
放点马蹄。

Local Fresh

UNIT FIVE

Ingredients

Look and say

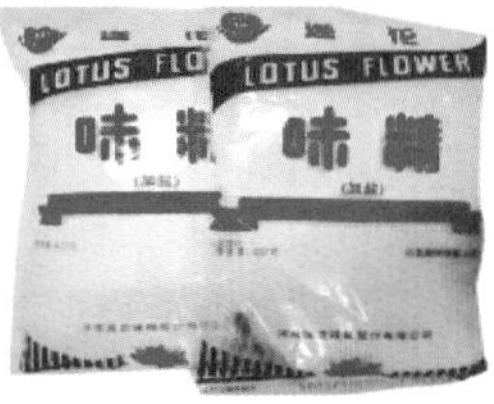

1.________________

2.________________

3.________________

4.________________

5.________________

6.________________

7.________________

8.________________

9.________________

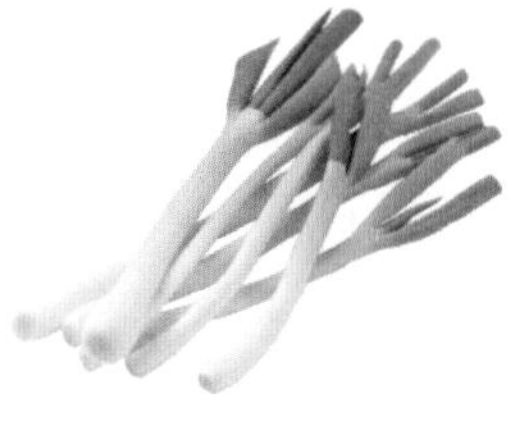

10.________________

11.________________

12.________________

Task I Find out the suitable pictures for the given words

chili	rapeseed oil	salad oil	scallion
ginger	garlic	MSG	salt
soy sauce	vinegar	honey	aniseed

Speak out

1. Learn the numbers

0 1 2 3 4 5 6

zero one two three four five six

7 8 9 10

seven eight nine ten

11 12 13 14 15 16

eleven twelve thirteen fourteen fifteen sixteen

17 18 19 20

seventeen eighteen nineteen twenty

21 30 40 50 60

twenty-one thirty forty fifty sixty

70 80 90 100

seventy eighty ninety one hundred

Task II Recite the numbers and read aloud

2. Tell the following telephone numbers and room numbers

Telephone numbers:

(1) ______110______ (2) ______120______

(3) ______4119270______ (4) ______3851226______

(5) ______12580______

Room numbers:

(1) ______2305______ (2) ______418______

(3) ______609______ (4) ______7231______

(5) ______8______

Task III Report the total amount of the following credit cards

$402

¥86

¥50

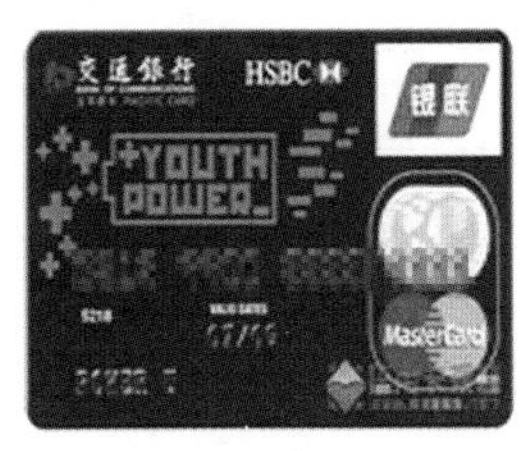

$100

£347

£900

3. Everyday English

- How much is it?
- How many are they in all?
- What's this for?
- It's for the service charge and the receipt.
- Here's your change.
- Welcome again next time.

Dialogues

Paying the bill

Dialogue I

—Waiter.

—Yes, sir.

—Can I have my bill, please?

—Here you are, sir.

—Can I sign for it?

—Certainly, sir. I hope you enjoyed your food.

—I really do.

—Thank you, sir. We hope to serve you again.

—Thank you and goodbye.

—Goodbye.

Task IV　Work in pairs

Fill in the blanks with correct words.

—Waiter.

—________, sir.

—Can I have my bill, please?

—________ you are, sir.

—Can I sign for it?

—Certainly, sir. I hope you ________ your food.

—I really do.

—Thank you, sir. We hope to ________ you again.

—Thank you and goodbye.

—________.

***Dialogue* Ⅱ**

—Waiter.

—Yes, sir.

—The bill, please.

—OK.

—How much?

—Two hundred and thirty yuan.

—What's this for?

—It's for the service charge.

—I see.

—Here's your change and the receipt.

—Thank you.

—My pleasure. Welcome again next time.

Task V　Work in pairs

Finish the dialogue with correct sentences.

—Waiter.

—________________________________

—The bill, please.

—________________________________

—How much is it ?

—________________________________

—What's that for?

—________________________________

—Oh, I see.

—________________________________

—Thank you very much.

—________________________________

Task VI Role-play

After the meal, the guests want to pay the bill. You come over and receive the money.

Useful sentences with *carve*

Carve cross lines on the meat.
Carve the turnip like a China rose.
Carve the tomato like a dahlia.
Carve the pumpkin like a dragon's head.
Carve the carrot like a lobster (herbaceous peony, tree peony, chrysanthemum flower).

Task VII Discussion

Discuss how to say the following sentences in English.
在鲤鱼上划上十字纹。
在西瓜上面雕一朵芍药花。
把洋葱刻成荷花。
把冬瓜刻成花篮。
把南瓜刻成龙头。

More words about ingredients

pepper

mustard

lard

sesame oil | jam | wild pepper

marmalade | cream | butter

alum | anchovy sauce | sweet sauce

olive oil | brown sugar | mayonnaise

peppermint | spices | cassia

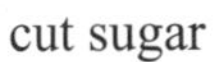
cut sugar

cheese

rock sugar

Task VIII Fill in the blanks with correct letters

____ili	rapes____d oil	salad ____l
scall____n	ging____	g____lic
M____G	s____t	s____ sauce
vineg____	pepp____	must____d
l____d	ses____me oil	j____m
marmal____de	cr____m	butt____
ch____se	al____m	h____ney
anchov____ sauce	sw____t sauce	ol____ve oil
w____ld pepper	br____n sugar	mayonn____se
pepperm____nt	anis____d	sp____ces
____nise	cut s____gar	ca____ia
r____ck sugar		

Recipe

Winter Stew

Preparation time: 2 hours (serve 4 ~ 6)

Ingredients:

4 onions

1 small celeriac

2 tsp ghee

4 cloves of garlic	1tsp gluten
2 leek	2 carrots
1 small turnip	1.1 litre vegetable stock
2 tsp yeast extract	black pepper

Procedures:

Preheat the oven at gas 3 (160 ℃). Wash and prepare the vegetables, peeling only the onions and celeriac. Brush an oven casserole dish with a little ghee in which to sauté the vegetables first or alternatively sauté in a separate frying pan. Crush the garlic, slice the onions and sauté them for 10 minutes. Add the gluten and continue to cook for a further 5 minutes, stirring occasionally, while you chop the remaining vegetables. Add all these to the pan and stir well. Transfer your oven casserole dish if you have been using a frying pan. Pour in the stock, add the seasoning and yeast extract. Stir again and cover with a lid. Place the casserole in oven and cook slowly for at least 1～2 hours.

Task IX　Questions about the recipe

1. How long do we need to prepare this dish?

2. How many people are the dish prepared for?

3. How long can the onions be cooked?

4. How long do we cook slowly after all the preparation is ready?

Task X Talk about the above dish according to the pictures

Assignment

1. Recite the numbers and learn to report the total amount of the credit cards.
2. Copy the words about ingredients and recite them.
3. Practise the dialogue with your partner.
4. Write out a short dialogue about paying the bill.
5. Make sentences with the word *carve*.
6. Translate the English recipe into Chinese.
7. Try to describe a Chinese recipe in English.

Self-evaluation

1. I can speak out the numbers and report the total amount of the credit cards. (　　)
2. I can pronounce and remember the words. (　　)
3. I can practise the dialogue. (　　)
4. I can use the word *carve* correctly. (　　)

5. I can understand the English recipe. (　　)
6. I can tell others the recipe in English. (　　)
7. My questions are ________________________________.

Word list for Unit Five

lard [lɑːd] 猪油
rapeseed oil [ˈreipˌsiːd ɔil] 菜油
salad oil [ˈsæləd ɔil] 色拉油
garlic [ˈɡɑːlik] 蒜
pepper [ˈpepə] 花椒
MSG [ˌem es ˈdʒiː] 味精
salt [sɔːlt] 盐
soy sauce [ˌsɔi ˈsɔːs] 酱油
vinegar [ˈvinigə] 醋
mustard [ˈmʌstəd] 芥茉
jam [dʒæm] 果酱
marmalade [ˈmɑːməleid] 橘子酱
cream [kriːm] 奶油
butter [ˈbʌtə] 黄油
cheese [tʃiːz] 奶酪
alum [ˈæləm] 明矾
honey [ˈhʌni] 蜂蜜
anchovy sauce [ˈæntʃəvi sɔːs] 鱼汁
sweet sauce [swiːt sɔːs] 甜酱
brown sugar [ˌbraun ˈʃugə] 红糖
mayonnaise [ˌmeiəˈneiz] 蛋黄酱
peppermint [ˈpepəmint] 薄荷
aniseed [ˈænəsiːd] 大料
spice [spais]香料
anise [ˈænis]大茴香
cut sugar [kʌt ˈʃugə] 方糖
rock sugar [rɔk ˈʃugə] 冰糖
cassia [ˈkæsiə] 桂皮
dahlia [ˈdeiliə] 天竺牡丹
herbaceous peony [həːˈbeiʃəs ˈpiːəni] 芍药花
cube [kjuːb] 立方体
tree peony [triː ˈpiːəni] 牡丹
chrysanthemum flower
[kriˈsænθəməm ˈflauə] 菊花
carp [kɑːp] 鲤鱼
lotus [ˈləutəs] 荷花
basket [ˈbɑːskit] 篮子
celeriac [səˈleriæk] 一种肥根芹菜
gluten [ˈɡluːtn] 面筋
alternatively [ɔːlˈtəːnətivli] 两者任选其一
further [ˈfəːðə] 进一步的
transfer [trænsˈfəː] 转移
litre [ˈliːtə] 公升
preheat [ˌpriːˈhiːt] 预热
casserole [ˈkæsərəul] 砂锅
oven [ˈʌvn] 火炉
clove [kləuv] 丁香
stock [stɔk] 原汤
yeast extract [ˈjiːst ekstrækt] 酵母粉
crush [krʌʃ] 压碎
ghee [ɡiː] 酥油
separate [ˈsepəreit] 隔离
continue [kənˈtinjuː] 继续
lid [lid] 盖子
remaining [riˈmeiniŋ] 剩余的
service [ˈsəːvis] 服务
sauté [ˈsəutei] 嫩炒
seasoning [ˈsiːzəniŋ] 调味品
sign [sain] 签字
receipt [riˈsiːt] 收据

Chinese for the useful sentences.
在肉上划上十字纹。
把萝卜刻成月季花。
把西红柿刻成天竺牡丹。
把南瓜刻成龙头。
把胡萝卜刻成龙虾（芍药花、牡丹、菊花）。

Salt Roasted Chicken

UNIT SIX

Nuts and Sweetmeats

Look and say

1.________________

2.________________

3.________________

4.________________

5.________________

6.________________

7.________________

8.________________

9.________________

10.________________

11.________________

12.________________

Task I Find out the suitable pictures for the given words

pistachio	chestnut	almond	Chinese date
walnut	peanut	raisins	confect
melon seeds	longan	fig	lotus nut

Speak out

1. Learn the expressions of time

A: What time is it? B: It's one o'clock.

A: What time is it? B: It's half past 8./It's eight thirty.

A: What time is it? B: It's a quarter past one./ It's one fifteen.

Task II Recite the words about time and read aloud

2. Tell and write the following time in English

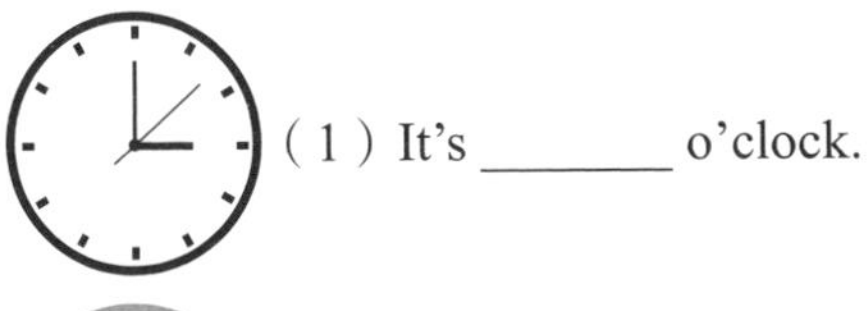

(1) It's ________ o'clock.

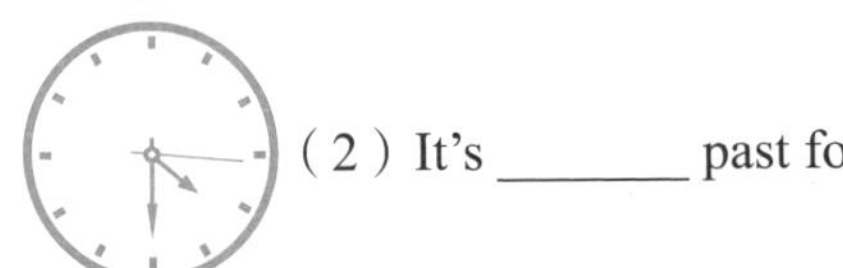

(2) It's ________ past four./It's four ________.

(3) It's _______ past twelve./It's twelve _______.

(4) It's _______ past ten./It's ten _______.

Task III Make a survey in English: find out when your partners have three meals each day

Questions	Names			
	Li Hong			
When do you have your breakfast?				
When do you have your lunch?				
When do you have your dinner?				

3. Everyday English

- Will you please tell me the daily service hours?
- When does your restaurant close?
- When does it open?
- Lunch time is from 12:00 to 13:30.
- Many thanks.
- It's one of our services.

Dialogues

Dialogue* I *Reservation

—Chinese Restaurant, may I help you?

—What time do you open this evening?

—At 6:00, sir. And we close at 10:00.

—I'd like to reserve a table by the window for four, please.

—For what time, sir?

—Around 8:15.

—May I have your name, please, sir?

—Blake.

—A table by the window for four this evening at 8:15 for Mr Blake.

—That's right.

—Thank you, sir.

—Thank you, goodbye.

Task IV Work in pairs

Fill in the blanks with correct words.

—Chinese Restaurant, may I _______ you?

—What time do you open this evening?

—_______ 6:00, sir. And we _______ at 10:00.

—I'd like to reserve a table by the window for four, please.

—For what _______, sir?

—Around 8:15.

—May I have your _______, please, sir?

—Brown.

—A table by the _______ for four this evening at _______ for Mr. Brown.

—That's right.

—Thank you, _______.

—Thank you, goodbye.

Dialogue **II** *The service hours*

—Will you please tell me the daily service hours of the dinning room?

—Certainly, sir. The dinning room opens at 7:00 am every day. Breakfast is from 7:30 to 9:00 am. Lunch from 12:00 to 2:00 pm. And supper from 6:00 to 8:00 pm.

—When does it close?

—It closes at 8:00 pm on Monday, Tuesday, Thursday, and Friday. And it closes at 10:00 pm on Wednesday and Saturday.

—When does the bar open?

—From 4:00 pm till midnight every day.

—Many thanks.

—Not at all. It's one of our service.

Task V Work in pairs

Finish the dialogue with correct sentences.

—Excuse me, could you tell me the daily service hours of the bar?

—______________________________

—When does it close?

—______________________________

—When does the dinning room open?

—______________________________

—Thank you very much.

—______________________________

Task VI Role-play

A guest comes to your hotel. He wants to know the time your dinning room opens and closes. Please tell him.

Useful sentences with *keep*

Keep the pig's trotters.
Keep the geese's webs.
Keep the bear's paws.

Task VII Discussion

Discuss how to say the following sentences in English.
留着鸡翅。
留着鸭掌。
留着鱼皮。
留着羊蹄。

More words about nuts and sweetmeats

currant

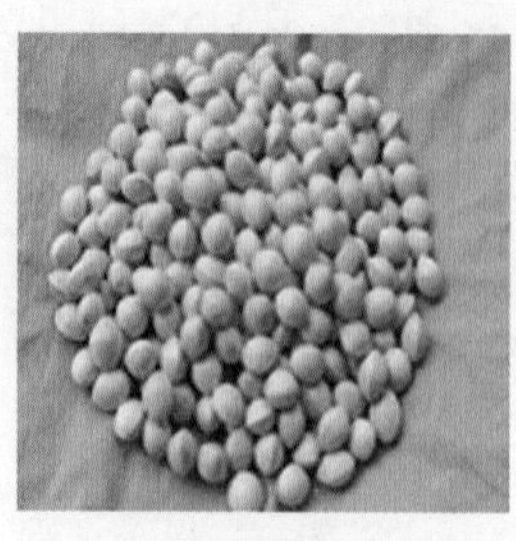
gingko

fruit salad

preserved apricot currant

persimmon cake

Task VIII Fill in the blanks with correct letters

lot____s nut	pistach____
____estnut	____mond
Chinese d____te	p____nut
r____sins	conf____ct
melon s____ds	long____n
f____g	c____rrant
g____ngko	fr____t salad
pres____ved apricot	p____simmon cake

Recipe

Diced Chicken with Chili Pepper

Ingredients:

400 g dark chicken meat

1 egg white	100 g peanuts
50 g lard	50 g cornstarch
15 g soy sauce	5 g vinegar
10 g white sugar	5 chilies

10 g wild pepper-corns

10 g each of slices chives, ginger and garlic

5 g MSG

Procedures:

1. Cut the chicken into 2 inches cubes and mix in the egg white. Skin the peanuts. Remove the seeds from chilies and cut into 1 inch pieces.

2. Heat the lard in a pan. Add peanuts, and fry till crisp. Remove and reheat the pan and stir-fry the chives, garlic, ginger, and wild pepper-corns. Add chilies and then the chicken cubes, sauté until the meat is mostly cooked. Add MSG, sugar, and cornstarch, and sprinkle with vinegar. Sauté a few minutes more. Mix in fried peanuts and remove to serving plate.

Task IX Answer the questions

1. How many kinds of ingredients are there in this dish?

2. How much vinegar is needed?

3. How to prepare peanuts?

4. When do you put the chicken cubes in the pan?

Task X Talk about the above dish according to the pictures

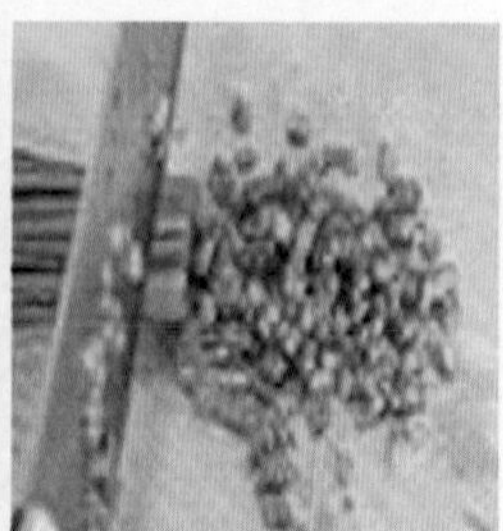

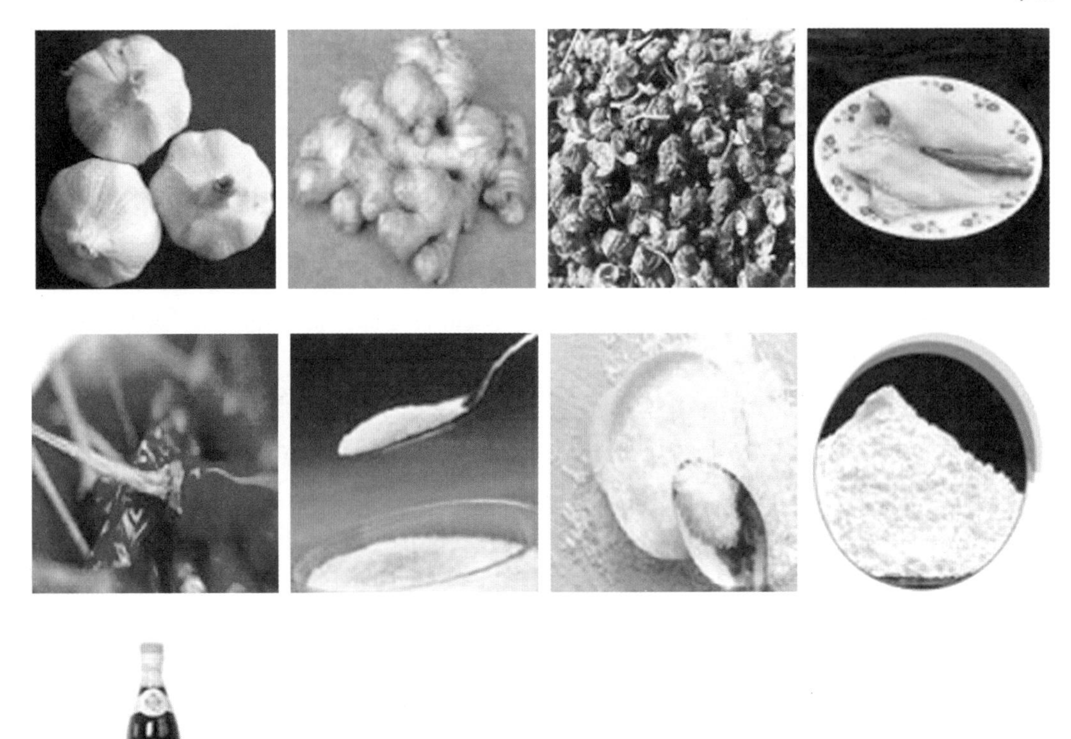

Assignment

1. Recite the expressions of time and learn to tell the time.
2. Copy the words on the nuts and sweetmeats and recite them.
3. Practise the dialogue with your partner.
4. Make sentences with the word *keep*.
5. Translate the English recipe into Chinese.
6. Write out a recipe in English.

Self-evaluation

1. I can tell the time according to the clock. (   )
2. I can pronounce and remember the words. (   )
3. I can practise the dialogue. (   )
4. I can use the word *keep* correctly. (   )
5. I can understand the recipe. (   )

6. I can tell others a recipe in English. (　　)
7. I can write out a recipe in English. (　　)
8. My questions are ________________________________.

Word list for Unit Six

lotus nut [ˈləutəs nʌt] 莲子
chestnut [ˈtʃesnʌt] 板栗
pumpkin seed [ˈpʌmpkin siːd] 南瓜子
preserved apricot [priˈzəːvd ˈæprikɔt] 杏脯
gingko [ˈɡiŋkəu] 银杏果
currant [ˈkʌrənt] 无核葡萄干
confect [kənˈfekt] 蜜饯
dice [dais] 将……切成丁（块）
web [web]（水禽等的）蹼
horizontal [ˌhɔriˈzɔntl] 水平的
rammish [ˈræmiʃ] 臭气强烈的
slash [slæʃ] 砍痕
depth [depθ] 深度
dining room [ˈdainiŋ ruːm] 餐厅
fishskin [ˈfiʃˌskin] 鱼皮
Chinese date [ˌtʃaiˈniːz deit] 枣
longan [ˈlɔŋɡən] 桂圆
pistachio [piˈstætʃiəu] 开心果
almond [ˈɑːmənd] 杏仁
walnut [ˈwɔːlnʌt] 核桃
fruit salad [fruːt ˈsæləd] 什锦水果
raisins [ˈreiznz] 葡萄干
watermelon seeds [ˈwɔːtəˈmelən siːdz] 西瓜子
persimmon cake [pəˈsimən keik] 柿饼
stir-fry [ˈstəː frai] 炒菜
skin [skin] 去皮
paw [pɔː] 爪子
trotter [ˈtrɔtə] 蹄
wing [wiŋ] 翅膀
midnight [ˈmidnait] 半夜
restaurant [ˈrestrərɔnt] 饭店
daily [ˈdeili] 日常的
peanut [ˈpiːnʌt] 花生
cube [kjuːb] 立方体

Chinese for the useful sentences.
留着猪蹄。
留着鹅掌。

Candied Banana

UNIT SEVEN 7

Drinks

Look and say

1.________________

2.________________

3.________________

4.________________

5.________________

6.________________

7.________________

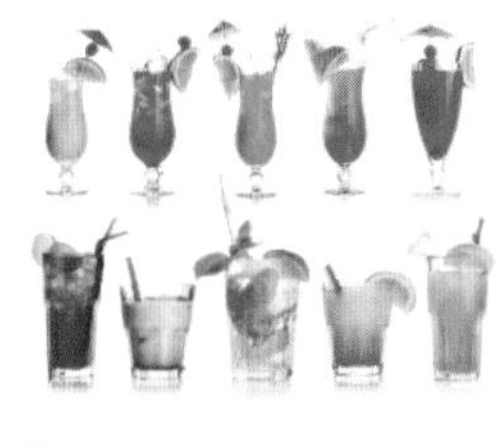

8.________________

9.________________

10.________________

11.________________

12.________________

Task I Find out the suitable pictures for the given words

mineral water	ice cream	liquor	grape wine
cocktail	fruit juice	beer	yellow rice wine
milk	sprint	green tea	coffee

Speak out

1. Learn the words about colours

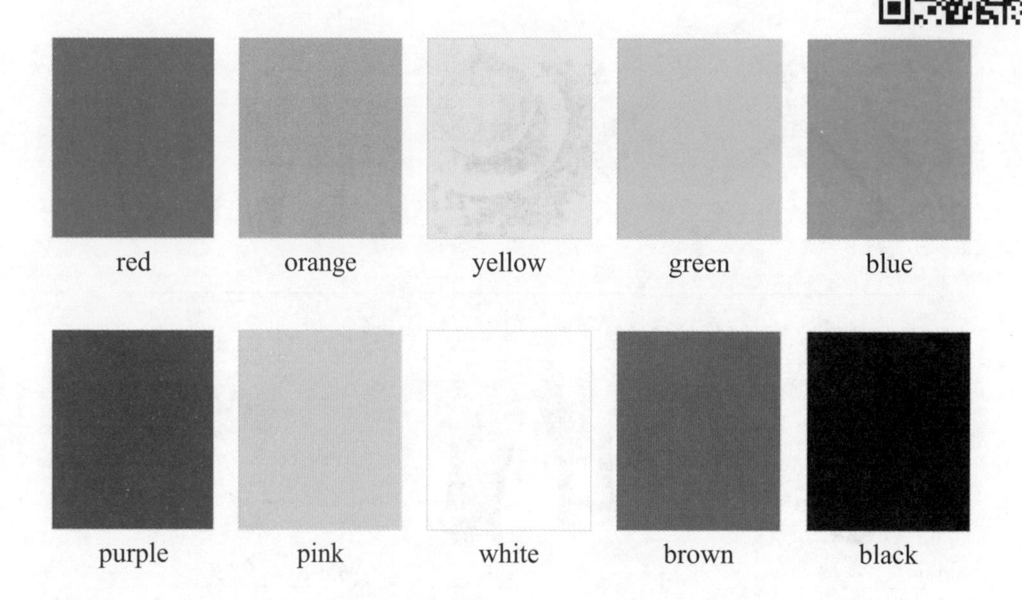

red orange yellow green blue

purple pink white brown black

Task II Recite the words about colours and read aloud

2. Make a short dialogue according to the example

black red yellow

Example: A: What colour is the coffee?

B: It's black.

Task III Fill in the blanks and read aloud the passage

Before you cross the road, you should watch the lights. When you see the ______ light, you must stop. When the ______ light is on, you may go or pass. The______ light means warning.

3. Everyday English

- When do you have breakfast?
- We have breakfast at about 9:00 in the morning.
- What do you usually have for your breakfast?
- Where do you work?
- Do you have vegetables every day?
- Do you have milk?
- I have noodles with eggs.
- We have super skills of cooking.

Dialogues

What do you have for breakfast?

Dialogue I

—When do you usually have breakfast?

—At 7:00.

—What do you usually have for breakfast?

—I have juice, cereal and tea.

—Do you have coffee for breakfast?

—Yes, I do.

—Do you have orange juice every day?

—No, I don't. Sometimes I have pineapple juice.

Task IV Work in pairs

Fill in the blanks with correct words.

—When do you usually have breakfast?

—_______ 7:00.

—What do you usually have for breakfast?

—I have _______, cereal and _______.

—Do you have coffee for breakfast?

—Yes, I _______.

—Do you have orange juice every day?

—No, I _______. Sometimes I have pineapple _______.

Dialogue II

—Excuse me , where do you work?

—I work in that hotel.

—Are you a cook there?

—Yes.

—How many cooks are there in your kitchen?

—There are twenty.

—When do you have breakfast?

—About nine.

—What do you usually have for your breakfast?

—We usually have noodles.

—Do you have milk?
—No. Sometimes we have noodles with eggs.
—Can you prepare Western food?
—Yes. We have super skills of cooking.
—Thank you very much.
—You are welcome.

Task V Work in pairs

Finish the dialogue with correct sentences.

—Excuse me, where do you work?
—______________________________
—Are you a cook there?
—______________________________
—How many cooks are there in your hotel?
—______________________________
—When do you have supper?
—______________________________
—What do you usually have for your supper?
—______________________________
—Do you have milk?
—______________________________
—Can you prepare Western food?
—______________________________
—Thank you very much.
—______________________________

Task VI Role-play

Your boss comes to your kitchen and asks some questions about the food you have every day. Tell him what you have for dinner every day.

Useful sentences with *sprinkle*

Sprinkle salt on the dish.
Sprinkle some chili on the mutton.
Sprinkle peppered salt.
Sprinkle some sesame oil on the cold dish.
Sprinkle some mustard oil on the seafood.

Task VII Discussion

Discuss how to say the following sentences in English.

在菜上撒点糖。

淋些花生油。

在汤上撒些胡椒粉。

淋点色拉油。

More words about drinks

black coffee

white coffee

champagne

cocoa

lemonade

beer

stout

draught beer

brandy

sherry

vermouth

vodka

spirits

gin

whisky

soda water

sprint

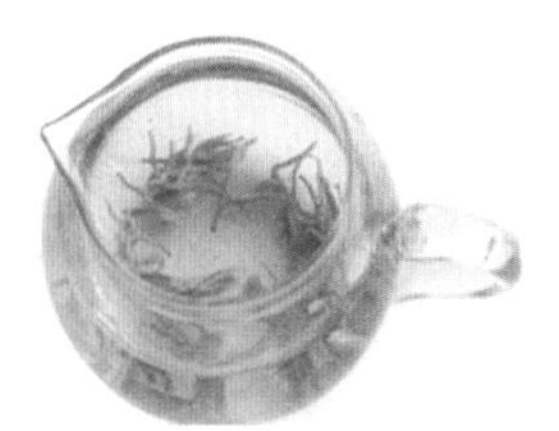

green tea

black tea

jasmine tea

scented tea

Task VIII Fill in the blanks with correct letters

m____neral water	ice cr____m	liqu____
grape w____	____ampagne	cockt____l
fruit j____ce	dr____t beer	yell____rice wine
br____ndy	black coff____	wh____te coffee
coc____	lem____nade	b____
st____t	____erry	v ____mouth
vodk____	spir____ts	g____n
whisk____	s____da water	Spr____te
green t____	bl____ck tea	j____smine tea
sc____nted tea		

Recipe

Champagne Punch

Ingredients:

- 2,750 mL bottle sparkling wine
- 1,750 mL bottle white dinner wine
- 6 whole cloves
- 4 cinnamon sticks
- 1 lemon
- 110 g granulated sugar (or to taste)
- 475 mL orange juice
- 240 mL pineapple juice
- ice block

Method:

1. In a saucepan combine white wine, cloves, cinnamon.
2. Squeeze in lemon juice, and then add the rest of the lemon.
3. Bring the mixture to a boil.
4. Lower heat and simmer for 10 minutes.
5. Strain and cool completely.
6. Place ice in a punch bowl.
7. Pour in spice/wine mixture, sparkling wine, and juices. Stir gently and serve.

Task IX True or false

1. There are nine ingredients in the champagne punch. (　　)
2. There is no ice in the wine. (　　)
3. Put all the lemon juice in the wine. (　　)
4. Heat the juice and wine to boil. (　　)

Task X Talk about champagne punch according the pictures

Assignment

1. Recite the words on colours and learn to describe colours.
2. Copy the words on the drinks and recite them.
3. Practise the dialogue with your partner.
4. Write a short dialogue on how to make salad.
5. Make sentences with the word *sprinkle*.
6. Translate the English recipe into Chinese.
7. Try to describe a Chinese recipe in English.

Self-evaluation

1. I can describe the colours in English. (　　)
2. I can pronounce and remember the words. (　　)
3. I can practise the dialogue. (　　)
4. I can use the word *sprinkle* correctly. (　　)
5. I can understand the recipe. (　　)
6. I can tell others a recipe in English. (　　)
7. My questions are ______________________________.

Word list for Unit Seven

mineral water [ˈminərəl wɔːtə] 矿泉水
liquor [ˈlikə] 白酒
champagne [ʃæmˈpein] 香槟酒
ice cream [ˈais kriːm] 冰激凌
grape wine [greip wain] 葡萄酒
cocktail [ˈkɔkteil] 鸡尾酒

fruit juice [fruːt dʒuːs] 果汁
yellow rice wine [ˈjeləu rais wain] 黄酒
black coffee [blæk ˈkɔfi] 清咖啡
cocoa [ˈkəukəu] 可可
beer [biə] 啤酒
sherry [ˈʃeri] 雪利酒
vodka [ˈvɔdkə] 伏特加
gin [dʒin] 杜松子酒
soda water [ˈsəudə wɔːtə] 苏打水
green tea [ˌgriːn ˈtiː] 绿茶
jasmine tea [ˈdʒæzmin tiː] 茉莉花茶
rind [raind] 果皮
pith [piθ] 木髓
grate [greit] 磨碎
decorate [ˈdekəreit] 装饰
drain [drein] 榨汁
completely [kəmˈpliːtli] 完全地
Western food [ˈwestən fuːd] 西餐
sparkling [ˈspɑːkliŋ]（欧洲）胡瓜鱼
cloves [kləuvz] 丁香
sticks [stiks] 木棍
taste [teist] 尝
saucepan [ˈsɔːspən] 平底锅
squeeze [skwiːz] 挤、榨、捏
lower [ˈləuə] 降低
sprinkle [ˈspriŋkl] 撒；淋
draught beer [dræft biə] 生啤酒
brandy [ˈbrændi] 白兰地
white coffee [wait ˈkɔfi] 牛奶咖啡
lemonade [ˌleməˈneid] 柠檬水
stout [staut] 黑啤酒
vermouth [vəˈmuːθ] 味美思
spirit [ˈspirit] 烈酒、酒精
whisky [ˈwiski] 威士忌
Sprite [sprait] 雪碧
black tea [blæk tiː] 红茶
scented tea [ˈsentid tiː] 花茶
refrigerate [riˈfridʒəreit] 冷藏、冷冻
arrange [əˈreindʒ] 整理
sprig [sprig] 枝状装饰图案
spiced salt [spaist sɔːlt] 椒盐
skill [skil] 技能
cereal [ˈsiəriəl] 谷类食品
punch [pʌntʃ] 潘趣酒
cinnamon [ˈsinəmən] 肉桂
granulated [ˈgrænjuleitid] 颗粒状的
ice block [ais blɔk] 冰块
combine [kəmˈbain] 混合
rest [rest] 剩下部分
simmer [ˈsimə] 炖、慢煮

Chinese for the useful sentences.
在菜上撒盐。
在羊肉上撒些辣椒。
撒椒盐。
在凉菜上淋些芝麻油。
在海鲜上淋点芥末油。

Beer Duck

UNIT EIGHT

Seafood

Look and say

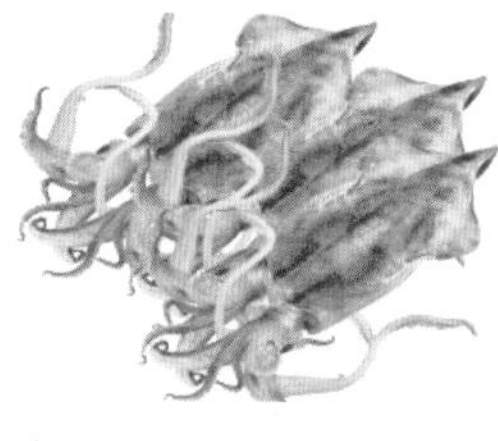

1.________

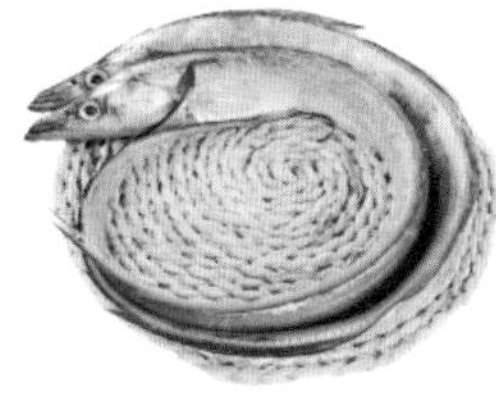

2.________

3.________

4.________

5.________

6.________

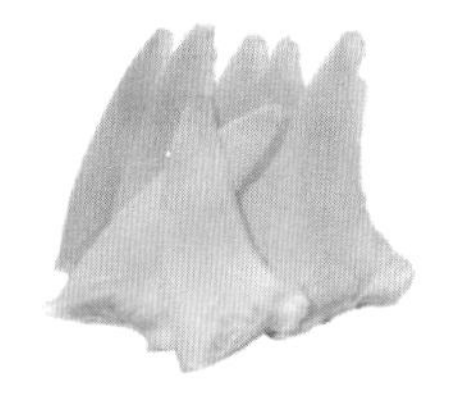

7.________

8.________

9.________

10.________

11.________

12.________

Task I Find out the suitable pictures for the given words

grouper	shrimp	oyster	hairtail fish
shark's fin	lobster	abalone	sea cucumber
eel	crab	squid	bass

Dialogue

Thai Prawn Soup and Chrysanthemum Fish

—The soup was hot and delicious. What is the name of the soup?

—It is called Thai Prawn Soup.

—What is it made of?

—It is made of prawns, small mushrooms, purple onions, lemon grass and scallions. It is then seasoned with fish sauce, chili pepper and lime juice, sprinkle with chopped coriander leaves and spring onions.

—It sounds appetizing, doesn't it?

—Yes, it is also very nutritious.

—And I've heard of Chrysanthemum Fish. What's the feature of this dish?

—It's a golden yellow colour dish and tastes crisp and tender. A very high cooking skill will be needed when cooking. Just right temperature control is very important, and it must be eaten immediately when cooked.

—Is that so? How does the chef prepare it?

—Slice a bass or a tilapia into fillets. Cut through the center of the fillets starting at one of the long sides. Do not cut all the way through the fillets: the idea is to "butterfly" it, so it opens up like a book. Then pour the oil into a pan and heat to 190 ℃, hold the fish strips by the uncut ends and slowly lower the cut end into the oil. Gently move the fish constantly as you lower it, so the cut ends begin to curl. When the ends begin to curl, release the fish into the oil. Cook until golden brown. At last, spoon the sauce over the fish.

—Thank you very much.

—Not at all.

Task II Finish the dialogue with suitable sentences

Thai Prawn Soup and Chrysanthemum Fish

—The soup is hot and delicious. What is the name of the soup?

—________________________________

—What is it made of ?

—________________________________

—It sounds appetizing, doesn't it?

—________________________________

—And I've heard of Chrysanthemum Fish. What's the feature of this dish?

—________________________________

—Is that so? How does the chef prepare it?

—________________________________

—Thank you very much.

—Not at all.

Task III Role-play

A guest comes to order a Chrysanthemum Fish. He wants to know how you prepare it. Tell him how you make the dish.

Useful sentences

Soak abalone in the warm water.
Scale the carp, wash it clean and remove entrails.
Wash the shrimp clean, chop off head and tail.
Fry the scallions and squid quickly and slightly.
Wash the lobster clean and carve it into shape of pine cones.
Slice sea cucumber diagonally into fillets and put it in a tray.

Task IV Discussion

Discuss how to say the sentences in English.
用温水浸泡鱼肚一天。
黄花鱼刮鳞，去掉头尾。
把带鱼放入油中略微爆炒一下。
将鲈鱼切成斜刀片，放入碟中。

More words about seafood

cob

mandarin fish

yellow croaker

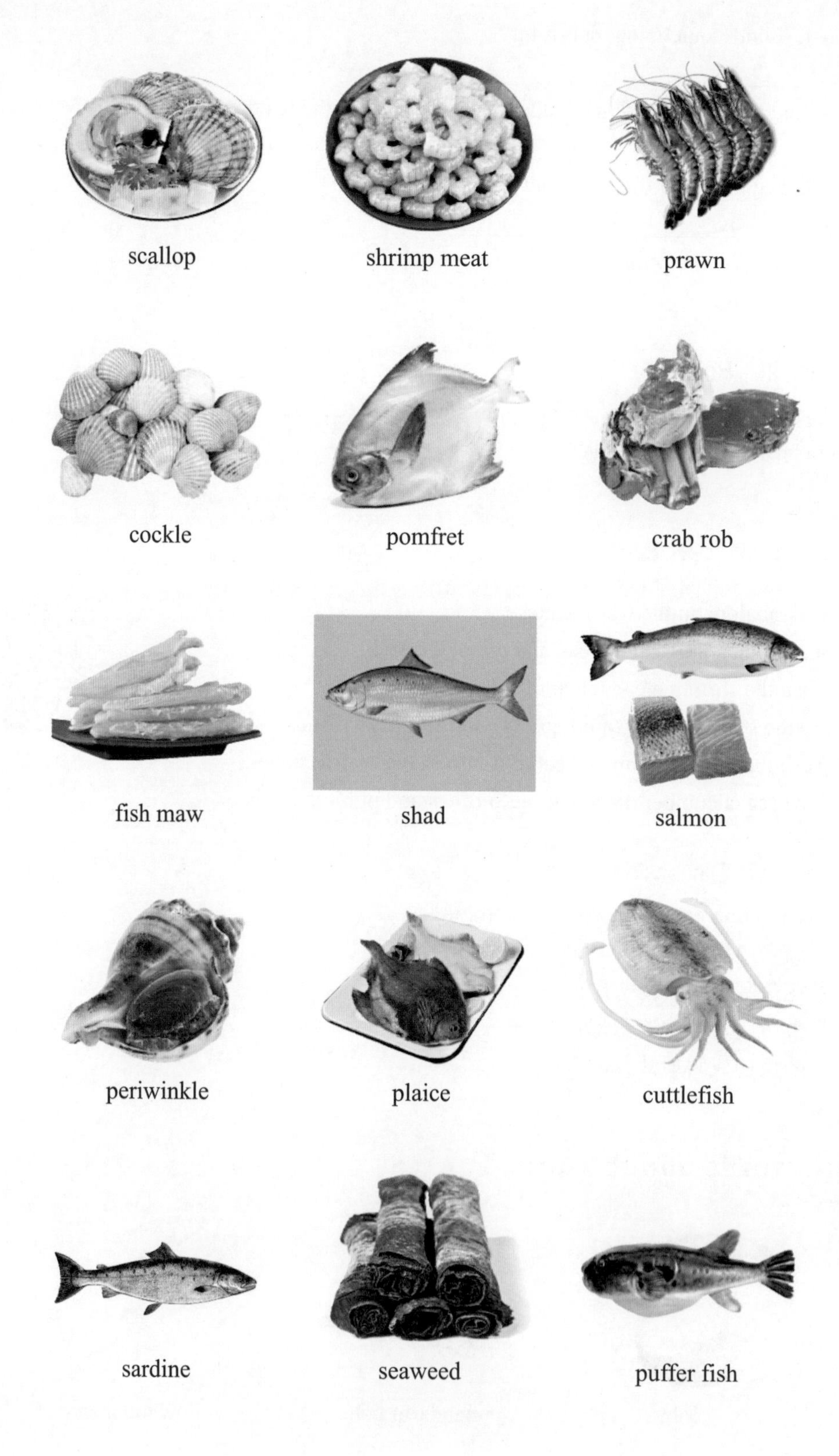
scallop
shrimp meat
prawn
cockle
pomfret
crab rob
fish maw
shad
salmon
periwinkle
plaice
cuttlefish
sardine
seaweed
puffer fish

tilapia

jelly fish skin

caviar

tuna

Task V Fill in the blanks with correct letters

t____tle
sh____k's f____n
____l
pl____ce
tun____
sc____llop
ab____lone
m____ndarin fish
c____ttl____f____sh
shrimp m____t
s____ cucumber
bird's n____st

p____mfret
b____ss
lobst____
____yst____r
b____ss
s____lm____n
sh____d
p____mfr____t
j____lly fish
gr____per
seaw____d
sq____d

shr____mp
s____dine
cr____b
c____v____r
pr____n
t____l____pia
y____ll____w cr____ker
h____tail fish
p____ff____r fish
per____winkle
fish m____w
cr____b r____b

Recipe

Marinated Grilled Shrimp

Ingredients:

- 3 cloves garlic, minced
- 1/3 cup olive oil
- 1/4 cup tomato sauce
- 2 tablespoons red wine vinegar
- 2 tablespoons chopped fresh basil
- 1/2 teaspoon salt
- 1/4 teaspoon cayenne pepper
- 2 pounds fresh shrimp, peeled and deveined
- skewers

Directions:

1. In a large bowl, stir together the garlic, olive oil, tomato sauce, and red wine vinegar. Season with basil, salt, and cayenne pepper. Add shrimp to the bowl, and stir until evenly coated. Cover, and refrigerate for 30 minutes to 1 hour, stirring once or twice.
2. Preheat grill for medium heat. Thread shrimp onto skewers, piercing once near the tail and once near the head. Discard marinade.
3. Lightly oil grill grate. Cook shrimp on preheated grill for 2 to 3 minutes per side, or until opaque.

Task VI True or false

1. Put the shrimp into a bowl and boil for a while. (　　)
2. The shrimp should be cooked over high heat all the time. (　　)
3. We should put the shrimp into the bowl with the mixture of garlic, tomato sauce and red wine vinegar. (　　)
4. Stir-fry the shrimp on the grill. (　　)
5. Cook shrimp on preheated grill for 2 to 3 minutes per side. (　　)

Task VII Talk about the above dish according to the pictures

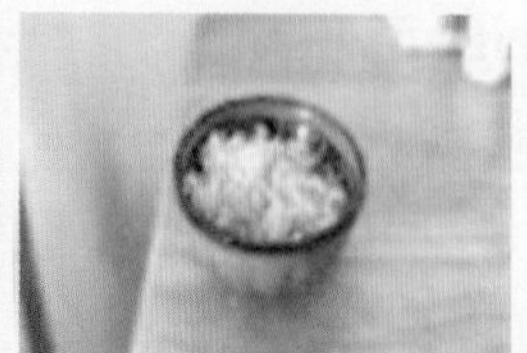

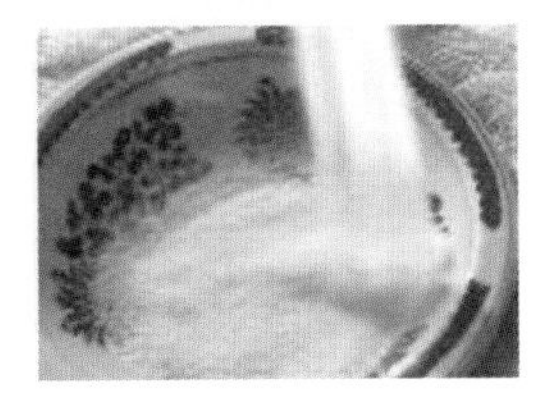

Assignment

1. Copy the words and recite them.
2. Make sentences with the words *soak*, *fry*, *scale, slice* and *remove*.
3. Recite the dialogue with your partner.
4. Write out a dialogue about how to make a seafood dish.
5. Translate the recipe into Chinese.
6. Write out a seafood recipe in English.

Self-evaluation

1. I can pronounce and remember the words. (　　)
2. I can understand the recipe. (　　)
3. I can practise the dialogue. (　　)
4. I can use the words *soak*, *fry*, *scale, slice* and *remove* correctly. (　　)
5. I can tell others how to make a seafood dish in English. (　　)
6. I can write out a seafood recipe in English. (　　)
7. My questions are______________________________.

Word list for Unit Eight

crab [kræb] 蟹
oyster [ˈɔistə] 牡蛎
squid [skwid] 鱿鱼
abalone [ˌæbəˈləuni] 鲍鱼
sea cucumber [ˌsiː ˈkjuːkʌmbə] 海参
yellow croaker [ˈjeləu ˈkrəukə] 黄花鱼
shrimp [ʃrimp] 小虾
mandarin fish [ˈmændərin fiʃ] 鳜鱼
lobster [ˈlɔbstə] 龙虾
hairtail fish ['hɛəteil fiʃ] 带鱼
turtle [ˈtəːtl] 甲鱼
cod [kɔd] 鳕鱼

bass [bæs] 鲈鱼
scallop [ˈskæləp] 扇贝
prawn [prɔːn] 明虾
silver carp [ˈsilvə kɑːp] 鲳鱼
crab rob [kræb rɔb] 蟹黄
shad [ʃæd] 鲥鱼
salmon [ˈsæmən] 三文鱼
cuttlefish [ˈkʌtlfiʃ] 墨鱼
seaweed [ˈsiːwiːd] 海带
tilapia [tiˈleipiə] 罗非鱼
bird's nest [bəːdz nest] 燕窝
lemon grass [ˈlemən grɑːs] 柠檬草
coriander（= parsley）[ˌkɔːriˈændə] 香菜
chrysanthemum fish [kriˈsænθəməm fiʃ] 菊花鱼
fresh [freʃ] 新鲜的
peeled [piːld] 去皮的
skewer [ˈskjuːə] 串住
refrigerate [riˈfridʒəreit] 使冷却
grill [gril] 烤架
piercing [ˈpiəsiŋ] 刺穿的
marinade [ˌmæriˈneid] 腌泡汁
grate [greit] 磨擦
soak [səuk] 浸泡
plaice [pleis] 比目鱼
shrimp meat [ʃrimp miːt] 虾仁
cockle [ˈkɔkl] 鸟蛤
shark's fin [ʃɑːks fin] 鱼翅
fish maw [fiʃ mɔː] 鱼肚
grouper [ˈgruːpə] 石斑
periwinkle [ˈperiwiŋkl] 海螺
sardine [ˌsɑːˈdiːn] 沙丁鱼
puffer fish [ˈpʌfə fiʃ] 河豚
jelly fish [ˈdʒeli fiʃ] 海蜇
caviar [ˈkæviɑː] 鱼子酱
lime [laim] 酸橙
tuna [ˈtjuːnə] 金枪鱼
chopped [tʃɔpt] 斩碎的
cayenne pepper [keiˈen ˈpepə] （红）辣椒粉
deveined [diveind] 除去（虾）的背部血管的
preheat [ˌpriːˈhiːt] 预热
thread [θred] 使……穿过
discard [diˈskɑːd] 丢弃
lightly [ˈlaitli] 少量地
opaque [əuˈpeik] 不透明的
scale [skeil] 刮鳞

Chinese for the useful sentences.
用温水浸泡鲍鱼。
鲤鱼刮鳞，洗净，去掉内脏。
把虾洗净，去掉头尾。
将葱和鱿鱼略微爆炒一下。
洗净龙虾肉，然后切成松子状。
将海参切成斜刀片，放入盘中。

UNIT NINE

Tableware and Cooking Tools

Abalone

Look and say

1.________________

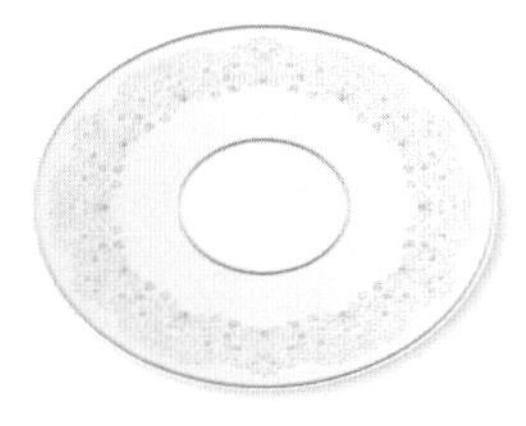

2.________________

3.________________

4.________________

5.________________

6.________________

7.________________

8.________________

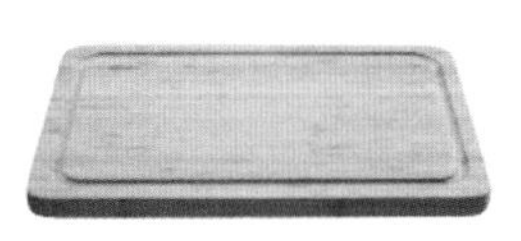

9.________________

10.________________

11.________________

12.________________

Task I Find out the suitable pictures for the given words

pan	plate	bowl	chopsticks
cup	knife	fork	tablespoon
chopping board	stove	basin	bucket

Speak out

Everyday English

- Do you serve Western-style food?
- Can you recommend some dishes to me?
- Do you have any chicken dishes?
- That dish is terrible.
- The stir-fried bean sprouts are very delicious.
- Would you like some drink?
- I'll take that one.
- How about Beijing duck?

Dialogue

Recommending dishes

—Do you serve Western-style food?

—I'm sorry, sir. We only serve Chinese food.

—Can you recommend some dishes to me?

—Certainly. The roast beef is very good today. Besides, the sweet and sour fish, and stir-fried bean sprouts are also very delicious.

—I don't like fish. Do you have any chicken dishes?

—Yes, sir. How about the diced chicken with fresh mushrooms?

—Good. I'll take these dishes.

—Would you like some drink?

—A bottle of beer and a glass of red wine.

—Anything else?

—No, thanks.

Task II Finish the dialogue with suitable sentences

—Do you serve Sichuan food or Cantonese food?

—________________________________

—Can you recommend your specialities to me?

—____________________

—I don't like beef. Do you have any mutton dishes?

—____________________

—Good. I'll take these dishes.

—____________________

—A bottle of beer and a glass of red wine.

—____________________

—No, thanks.

Task III Role-play

One of your friends comes to have dinner. He wants to have the specialities in your restaurant. You recommend them to your friend.

Useful sentences with *garnish*

Garnish the dish with cucumber slices.
Garnish the dish with vegetable leaves.
Garnish the dish with four boiled egg halves.
Garnish the dish with cherry.
Garnish the dish with carrot carved like a flower.

Task IV Discussion

Discuss how to say the sentences in English.
用西兰花装饰菜。
用香菜配菜。
用四根切成片的黄瓜配菜。
用萝卜雕成花装饰菜。
用草莓装饰蛋糕。

More word about tableware and cooking tools

coffee pot　　spoon　　tray

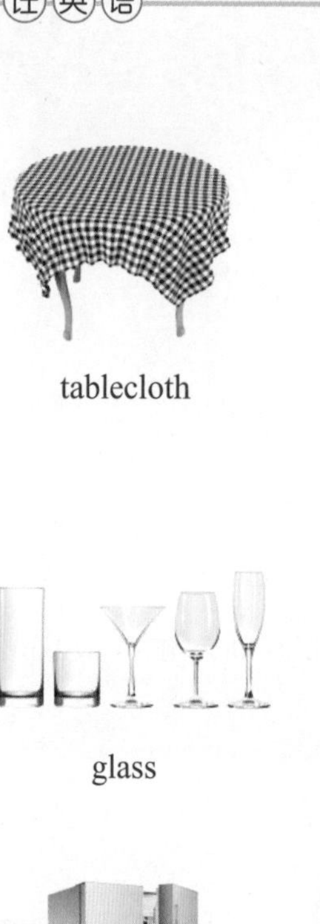

tablecloth

napkin

ash-tray

glass

tooth pick

clay pot

refrigerator

mug

goblet

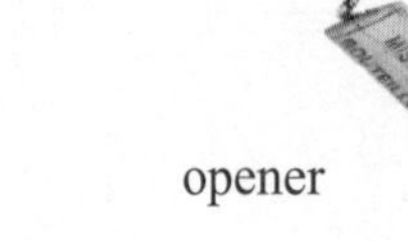

opener

toaster

egg-beater

meat-grinder

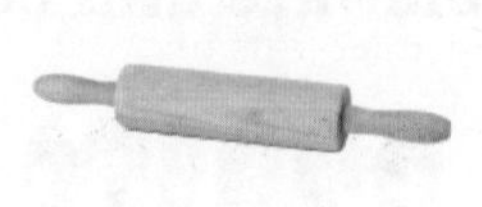

rolling pin

frying pan

stewing pan

pressure cooker

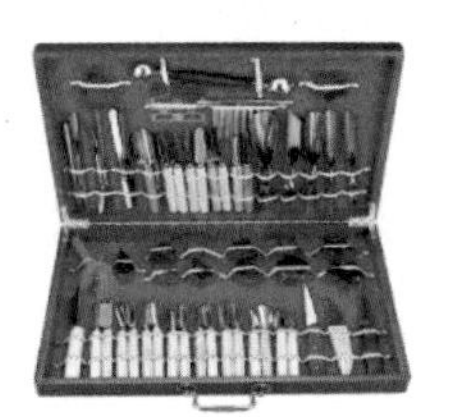
carving knife

Task V Fill in the blanks with correct letters

sp____n	pl____te	b____l
chopst____cks	c____p	kn____fe
f____k	table sp____n	chopping b____d
st____ve	coffee p____t	tr____ table
cl____th	n____pkin	a____-tray
gl____ss	t____th pick	cl____ pot
p____n	refrigerat____	m____g
gob____t	open____	t____ster
mix____	egg-b____ter	meat-grind____
r____lling pin	fr____ing pan	b____cket
st____ing pan	press____ cooker	c____ving knife
bas____n		

Instruction

Delicious Biscuit

High quality muesli with 40% fruit content

Ingredients:

Dried fruits: sultanas, banana chips (bananas, coconut oil, sugar, honey, flavouring), pieces of apricot, plum, peach, apple date, fig, pear, strawberry, raspberry. Whole wheat flakes, whole oat flakes, cornflakes (maize, sugar, salt, barley malt extract). Whole rye flakes, apricot fruit powder (malt dextrin, apricot). Granadilla fruit powder (malt dextrin, granadila). Flavouring.

Nutritive value:

Energy	1,446 kJ (342) kcal	17%
Protein	7.7 g	13%
Fat	5.4 g	9%
Carbohydrate	65.6 g	22%

May contain traces of peanuts and other nuts (shell fruits).

The analytical values are subject to the usual biological fluctuations.

Keep in cook, dry place/best before: see on top.

Task VI True or false

1. This is a kind of dessert. ( )
2. This food contains 10 dried fruits. ( )
3. It is very nutritive. ( )
4. There are some sugar, honey and powder in the food. ( )

Assignment

1. Copy the words on cooking tools and recite them.
2. Make sentences with the word *garnish*.
3. Practise the dialogue with your partner.
4. Translate the instruction into Chinese.
5. Write a recipe in English.

Self-evaluation

1. I can pronounce all the words. ( )
2. I can understand the instruction. ( )
3. I can practise the dialogue. ( )
4. I can use the word *garnish* correctly. ( )
5. I can use the goods correctly according to the instruction. ( )
6. My questions are ______________________________.

Word list for Unit Nine

pan [pæn] 锅
bowl [bəul] 碗
chopsticks [ˈtʃɔpstiks] 筷子
knife [naif] 刀
tablespoon [ˈteiblspuːn] 大汤匙
stove [stəuv] 灶
coffee pot [ˈkɔfi pɔt] 咖啡壶
tray [trei] 托盘
napkin [ˈnæpkin] 餐巾
glass [glɑːs] 玻璃杯
menu [ˈmenjuː] 菜单
mug [mʌg] 啤酒杯
opener [ˈəupnə] 开瓶器
mixer [ˈmiksə] 搅拌机
meat-grinder [miːtˈgraində] 绞肉机
frying pan [ˈfraiiŋ pæn] 煎锅
pressure cooker [ˈpreʃə kukə] 高压锅
delicious [diˈliʃəs] 美味的
content [ˈkɔntent] 内容、目录
sultanas [sʌlˈtɑːnəz] 小葡萄干
flavouring [ˈfleivəriŋ] 调味品
flakes [fleiks] 鳞片
cornflakes [ˈkɔːnfleiks] 玉米片
barley [ˈbɑːli] 大麦
malt [mɔːlt] 麦芽
malt dextrin [mɔːlt ˈdekstrin] 麦芽糊精
carbohydrate [ˌkɑːbəuˈhaidreit] 碳水化合物
nutritive [ˈnjuːtritiv] 营养物
kcal (=kilocalorie) [ˈkiləˌkæləri] 千卡
traces [ˈtreisiz] 痕迹
goods [gudz] 商品
recommend [ˌrekəˈmend] 推荐
biological fluctuations [ˌbaiəˈlɔdʒikl, flʌktjuˈeiʃənz] 生物波动

plate [pleit] 盘
bucket [ˈbʌkit] 水桶
cup [kʌp] 茶杯
fork [fɔːk] 叉
chopping board [ˈtʃɔpiŋ bɔːd] 切菜板
basin [ˈbeisn] 盆
teaspoon [tiːspuːn] 茶匙
tablecloth [ˈteiblklɔθ] 桌布
ash-tray [æʃ trei] 烟灰缸
tooth pick [tuːθ pik] 牙签
clay pot [klei pɔt] 陶壶
refrigerator [riˈfridʒəreitə] 电冰箱
goblet [ˈgɔblət] 高脚酒杯
toaster [ˈtəustə] 电烤面包器
egg-beater [eg ˈbiːtə] 打蛋器
rolling pin [ˈrəuliŋ pin] 擀面杖
stewing pan [ˈstjuːiŋ pæn] 炖锅
carving knife [ˈkɑːviŋ naif] 雕刀
biscuit [ˈbiskit] 饼干
quality [ˈkwɔləti]品质、特性
chips [tʃips]炸土豆条
raspberry [ˈræzberi] 覆盆子（一种像草莓、黑莓的水果）
oat [əut] 燕麦
maize [meiz] 玉米
extract [ˈekstrækt] 榨出物、汁
rye [rai] 黑麦
granadilla [ˌgrænəˈdilə] 百香果
energy [ˈenədʒi] 能量
value [ˈvæljuː] 价值
analytical [ˌænəˈlitikl] 分析的
shell [ʃel] 贝壳、外形
muesli [ˈmjuːzli] 牛奶什锦
powder [ˈpaudə] 粉末

Chinese for the useful sentences.
用黄瓜片装饰菜。
用菜叶配菜。
用四个切成半的熟鸡蛋配菜。
用樱桃配菜。
用胡萝卜雕成花装饰菜。

Sugar Molded

UNIT TEN

Cooking Terms

Look and say

1.________________

2.________________

3.________________

4.________________

5.________________

6.________________

7.________________

8.________________

9.________________

10.________________

11.________________

12.________________

Task I　Find out the suitable pictures for the given words

fried	roasted	braised	boiled
stir-fried	rinsed	red-cooked	sauté
steamed	steeped	grilled	torefried

Dialogue

Sweet and sour dishes

—Is it true that Americans like sweet and sour dishes?

—It's true. They often order sweet and sour dishes.

—What are the most popular sweet and sour dishes?

—Sweet and sour spareribs, sweet and sour pork, sweet and sour fish.

—How does the chef prepare the sweet and sour spareribs?

—The spareribs are cut up into small pieces, breaded and deep-fried before they are blended into a sweet and sour sauce of tomato, green pepper and pineapple chunks.

—What is the sweet and sour fish made of?

—The whole fish is first gutted and scaled and then breaded and fried in deep fat until golden brown and crispy. It is then masked with the sweet and sour sauce.

Task II　Finish the dialogue with suitable sentences

—It is said that sweet and sour fish is very delicious. Is that true?

—________________________________

—What are the main ingredients for this dish?

—________________________________

—How much sugar and vinegar do we need for that?

—________________________________

—How does the chef prepare the sweet and sour fish?

—________________________________

Task III　Role-play

Discuss how to prepare Diced Chicken with Peanuts and make a dialogue.

Useful sentences with *split*

Split the scallions.

Split the fish from belly.

Split the chicken from spine.

Split the prawn from the back.

Split them into two pieces.

Task IV Discussion

Discuss how to say the sentences in English.

把蒜切开。

把鱼从肚子开膛。

把牛从脊骨剖开。

把鸡从背部剖开。

把姜切成两片。

More word about cooking terms

turn over | quick-boil grilled | smoke

iced | salted | pickled

burned | carved | ground

mashed | stir | bake

sweet
sour
salty
bitter
hot
fresh
stale
raw
ripe
dried
hard
soft
crisp
delicious
fat

Task V Fill the blanks with correct letters

fr____d	t____sted	br____sed
b____led	st____-fried	r____nsed
red-c____ked	s____té	st____med
gr____lled	t____n over	____ick-boilng
gr____lled	sm____ked	torefr____d
ic____d	s____ted	pi____led
b____ned	c____ved	gr____nd
ma____ed	st____	b____ke

sw____t	s____r	s____ty
bitt____	h____t	fr____sh
st____le	r____	r____pe
dr____d	h____d	s____ft
cr____sp	delic____us	f____t
l____n	tend____	t____gh
med____m	under-d____ne	ov____-done
w____ll-done	h____t	c____ld

Instruction

Nice Soy Milk

Low sugar with calcium

Goodness of drink

Low sugar with calcium:

Only contains 2.5 g sugar per 100 mL

At least 25% lower in calories compared to regular soy milk.

6.3 g of soya protein per serving

The goodness of Low Sugar with Calcium:

Low-saturated fat

A cholesterol free food

A lactose free food

No preservatives

No artificial flavouring

Suitable for vegetarian

NUTRITION INFORMATION

energy	357 kJ (85kcal)	protein	69.3 g
total fat	3.3 g	monounsaturated fat	0.8 g
polyunsaturated fat	2.0 g	oemga3 fatty acids	12 mg
omega6 fatty acids	1.7 g	saturated fat	0.5 g

trans fatty acid	0 g	cholesterol	0 mg
carbohydrate	7.3 g	total sugars	6.3 g
lactose	0 g	galactose	0 g
dietary fibre	1.0 g	sodium	25 mg
Vitamin A	130 μg	Vitamin B_2	0.5 mg
Vitamin B_{12}	1.0 μg	Vitamin D_3	1.3 μg
calcium	168 mg		

Ingredients:

Soyabean extract, water, cane sugar, dietary fibre, calcium, corn oil, mineral salts and vitamins (vitamin B_2, vitamin B_1, vitamin A, vitamin D_3, vitamin B_{12})

Task VI Answer the questions

1. How much sugar is there per 100 mL soy milk?

2. Is there any cholesterol in the drink?

3. The vitamins are very rich, aren't they?

4. What kind of people is the soy milk suitable for?

Assignment

1. Copy the words on cooking terms and recite them.
2. Make sentences with the word *split*.
3. Practise the dialogue with your partner.
4. Translate the instruction into Chinese.
5. Write out a recipe in English.
6. Write out a dialogue on how to make a dish.

Self-evaluation

1. I can pronounce all the words. (　　)
2. I can understand the instruction. (　　)
3. I can make a dialogue on how to make a dish. (　　)
4. I can use the word *split* correctly. (　　)
5. I can use the goods correctly according to the instruction. (　　)
6. My questions are ______________________________.

Word list for Unit Ten

fried [fraid] 炸的
braised [breizd] 焖的
stir-fried [ˈstəː fraid] 炒的
red-cooked [ˈred kukd] 红烧的
steamed [stiːmd] 蒸的
turn over [təːn ˈəuvə] 两面煎
smoke [sməuk] 熏
iced [aist] 冰镇的，加冰块的
pickled [ˈpikld] 卤的
carved [kɑːvd] 切好的
mashed [mæʃt] 捣烂的
bake [beik] 烘、烤
sour [ˈsauə] 酸的
bitter [ˈbitə] 苦的
fresh [freʃ] 新鲜的
raw [rɔː] 生的
dried [draid] 干的
soft [sɔft] 软的
delicious [diˈliʃəs] 美味的
lean [liːn] 瘦的
rare [rɛə] 半生半熟的
medium [ˈmiːdiəm] 适中的
over-done [ˈəuvə dʌn] 过老
hot [hɔt] 热的，辣的
calcium [ˈkælsiəm] 钙
regular [ˈregjələ] 一般的
polyunsaturated [ˌpɔliʌnˈsætʃəreitid] 多不饱合的
roasted [ˈrəustid] 烤的
boiled [bɔild] 煮的
rinsed [rinst] 洗涮
vitamin [ˈvitəmin] 维生素
grilled [grild] 铁扒的
quick-boil [kwik bɔil] 氽，快煮的
torefried [tɔːfraid] 干炒的
salted [ˈsɔːltid] 腌的
burned [bəːnd] 烧焦的
ground [graund] 磨碎的
stir [stəː] 搅拌
sweet [swiːt] 甜的
salty [ˈsɔːlti] 咸的
stale [steil] 不新鲜的
ripe [raip] 熟的
hard [hɑːd] 硬的
split [split] 切开、剥开
fat [fæt] 肥的
tender [ˈtendə] 嫩的
tough [tʌf] 老的
under-done [ˈʌndə dʌn] 半生不熟的
well-done [wel dʌn] 全熟
cold [kəuld] 冷的，凉的
calories [ˈkæləriz]卡路里
compare [kəmˈpɛə] 对比

low-saturated [ləuˈsætʃəreitid] 低饱和的
lactose [ˈlæktəus] 乳糖
artificial [ˌɑːtiˈfiʃl] 人造的
suitable [ˈsuːtəbl] 适合的
fatty acids [ˈfæti ˈæsidz] 脂肪酸
trans [trænz] 反式的
dietary fibre [ˈdaiətəri ˈfaibə] 膳食纤维
monounsaturated [mɔnəuˌʌn'sætʃəreitid] 单不饱合的
cholesterol [kəˈlestərɔl] 胆固醇
preservatives [priˈzəːvətivz] 防腐剂
vegetarian [ˌvedʒəˈtɛəriən] 素食者
total [ˈtəutl] 全部的
saturated [ˈsætʃəreitid] 饱合的
galactoses [gəˈlæktəusis] 半乳糖
sodium [ˈsəudiəm] 钠
steeped [stiːpt] 泡的

Chinese for the useful sentences.
把葱切开。
把鱼从肚子开膛。
把鸡从脊椎骨剖开。
把虾从背部剖开。
把它们剖成两片。

Roast Suckling Pig

Supplementary Materials

A. Methods to Express the Names of Dishes

1. 主要烹调方法+主料+with +配料

e.g.

fried fish slices with tomato sauce	茄汁鱼片
steamed duck with mushrooms	口蘑蒸鸭
fried pork slices with onion sauce	葱汁肉片
roast pork chop with sour cabbage	酸菜烤猪排

2. 如果一个菜中有两个主料，可用“烹调方法+主料+and +主料”

e.g.

stewed crab meat and mushrooms	蘑菇蟹肉

3. 如果配料不突出，用“烹调方法+主料”即可

deep-fried fish slices	炸鱼片
steamed porgy	清蒸加吉鱼

4. 也可用“主料+with+配料”

e.g.

sea cucumber with crab meat	蟹肉海参
shark’s fin with chicken	鸡爪鱼翅
cucumber with sweet and sour sauce	糖醋黄瓜

5. 一些凉菜可根据其形象来译

e.g.

peacock cold dish	孔雀冷盘
flower basket	花篮
gold and silver prawns	金银大虾

6. 汤菜用“主料+soup”或“soup with +主料”, 也可用“主料+soup +with +配料”

e.g.

tomato soup	番茄汤
shredded chicken soup	鸡丝汤
soup with fish and shrimps	鱼虾汤

7. 也可根据风味和属性来译

e.g.

sour and sweet cabbage	糖醋白菜
cold and mixed meat	冷肉杂拌
sweet and sour fish slices	糖醋鱼块
cold ham	火腿

Practice:

Try to put the following into English.

酸辣面　　鸡蛋汤

回锅肉　　凉拌心舌

粉蒸肉　　樱桃蛋糕

B. Courtesy English

Greeting:

Good morning. Welcome to our hotel (restaurant). It's nice to see you again.

Farewell:

Mr. Lee, hope to see you again. Have a nice day.

Apology:

I'm sorry, sir (madam).

When a guest calls you:

Yes, sir. Can I help you?

When a guest asks for your help:

Certainly, sir (madam). I'll be right with you.

When a guest says "Thank you" to you:

It's my pleasure, sir (madam).

When a guest says "I'm sorry" to you:

That's all right.

When you need the guest wait for you:

Just a moment, please.

After the guest has waited for you:

I'm sorry to have kept you waiting, sir (madam).

When a guest is in a conversation:

Excuse me, sir (madam). I'm sorry to interrupt you, but…

When you can't understand what the guest has said:

I'm sorry, sir (madam). I didn't catch what you said. Would you please say it again?

C. English Names for Departments in Hotel

General Manager's Office　总经理办公室

Executive Office　行政办公室

Human Resources　人事资源部

Training Department　培训部

Accounting & Finance 财务部
Purchasing 采购部
Management Information Systems (MIS) 信息管理部
Sales & Marketing 市场销售部
Sales 销售部
Reservation 预订部
Inspired Events 灵感部
Food & Beverage 餐饮部
Marketing and Communication 市场公关部
Fukami-Japanese Restaurant 蓝日本餐厅
Mediterranean Fine Dinning Restaurant 地中海餐厅
Associate Dinning Room 员工餐厅
Associate Kitchen 员工厨房
Vie-All Day Dinning 全天候餐厅
In Room Dinning 送餐部
Link-Lobby Lounge 炫吧
Grand Ballroom-Banquet 宴会厅
Main Kitchen 主厨房
Western Kitchen 西厨房
Pastry 西饼房
Fruit Kitchen 水果房
Hygiene 实验室
Front Desk 前台接待
Housekeeping 客户部
Business Center 商务中心
Call Center 服务中心
La Spa 水疗中心
Laundry 洗衣房
Flower Shop 花房
Security 保安部
Parking Lot 停车场
Hair Saloon 美发沙龙
Theater Kitchen 开放厨房
Butchery 粗加工房
Cold Kitchen 凉菜厨房
Stewarding 管事部
Front Office 前厅部
Concierge 礼宾部
Lobby 大堂
Transportation 运输部
Health Club 健身中心
Floors 楼层
Public Area 公共区域
Clinic 医务室
Engineering 工程部
KIOSK 精品店
Le Shop 商店

参考文献

[1] 郑毅，杜纲. 酒店英语视听说[M]. 北京：外语教学与研究出版社，2010.
[2] 张艳红. 中餐烹饪英语[M]. 重庆：重庆大学出版社，2015.
[3] 张毅，贾颖丽. 烹饪厨房英语[M]. 2版. 重庆：重庆大学出版社，2021.